Agricultural Cooperatives

Agricultural Cooperatives

First Edition

Phil Kenkel

Oklahoma State University

Bassim Hamadeh, CEO and Publisher
Craig Lincoln, Project Editor
Rachel Kahn, Production Editor
Jess Estrella, Senior Graphic Designer
Laura Duncan, Licensing Coordinator
Natalie Piccotti, Director of Marketing
Kassie Graves, Senior Vice President, Editorial
Alia Bales, Director, Project Editorial and Production

Printed in the United States of America.

Contents

CHAPTER 1

The Cooperative Business Model

Introduction and Learning Objectives

This chapter introduces the cooperative business model and describes the basic ways in which cooperatives differ from other business forms. It provides a quick preview on the concepts of cooperative principles, cooperative profit distribution, and cooperative equity. The roles of the various participants in a cooperative are also discussed. Finally, the chapter discusses both traditional and nontraditional cooperative structures as well as overviewing the basic structures of investor-owned firms (IOFs) in the United States. After reading the chapter, you should have a basic concept of the nature of a cooperative business and have had an introduction to the upcoming chapters on nontraditional cooperatives, cooperative principles, and profit distribution and equity in cooperative firms.

What Is a Cooperative?

A *cooperative* is a unique form of private business organization that has been widely used for more than 100 years. Cooperatives operate in every sector of the economy, in both the United States and other market economies. In the United States, there are more than 30,000 cooperatives. You may have done business with a cooperative or purchased products such as Land O'Lakes butter or Ocean Spray cranberry juice without realizing that the product was manufactured by a cooperative. Although the cooperative form of business is widely used and successful, it is not well understood, even by those educated in economics and business. In this class, we will focus on cooperatives in the agricultural sector. Agricultural cooperatives are an important part of the U.S. agribusiness industry.

Farmers formed cooperatives because it allowed them to pool their resources and per business activities that they could not perform as economically on their own. This raises the concept of economies of scale, which is a common rationale for the formation of a cooperative. For example, the per-bushel cost of a cooperative grain elevator is typically much lower than that of on-the-farm storage. Producers can access inputs at a lower cost and market commodities at a higher cost through the scale economies of the cooperative. Forming a cooperative may also provide farmers with access to the market. For example, a dairy farmer cannot sell milk directly to supermarkets; however, as a member of a milk-processing cooperative, they have access to the retail milk market.

Why Should Learn About Cooperatives?

Why should you learn about the cooperative business model? As mentioned, cooperatives are an important part of the agricultural sector in the United States. If you plan a career in agriculture, it will be helpful to understand cooperatives. You may be employed by a cooperative, work for a firm competing with one, or buy or sell to one.

Of course, the people with the greatest need to understand cooperatives are the members of one. You or your family may now be a member of a cooperative, or you may become a member of an agricultural cooperative, a rural electric cooperative, a farm credit cooperative, a credit union, a local foods cooperative, or a cooperative in another sector. You will be better informed and prepared to make better personal and professional decisions involving cooperatives if you have a solid understanding of this unique form of business. Understanding how cooperatives are financed and governed and profits are generated and distributed will also improve your overall understanding of how a business operates.

More About the Nature of a Cooperative

In the United States, cooperatives are a special form of corporation. A simplified definition is that a cooperative is a corporation owned by its customers. More specifically, cooperatives are a special type of business that are owned by their users and from which the profits are distributed in proportion to use. One definition of a *cooperative* is as follows:

> A for-profit business that is owned and democratically controlled by the people who use its services and whose benefits are derived and distributed equitable on the basis of use.

Another definition that provides some insights into structure and operation is as follows:

> A for-profit corporation that distributes its revenues to its customer members in proportion to the volume of sales or purchased after deductions for operating expenses and reserves.

Cooperative businesses have a number of unique features that distinguish them from other business forms. Most U.S. businesses are owned by investors and distribute profits to those investors on the basis of ownership. Customers purchase goods and services from those businesses but are not owners, and the customers do not receive a share of the profits. Those businesses may take many legal forms, such as partnerships, limited liability companies, or C corporations. For simplicity, we will refer to those businesses as *IOFs*. Owners of an IOF control the firm, and their control—that is, their vote—is in proportion to their share of ownership in the firm. One individual can gain control over an IOF by owning more than 51% of the equity. With that background, the above definition suggest some of the unique differences between cooperatives and IOFs.

User-Owner

The first difference is that of user-owner. Cooperative businesses are owned by the individuals who use the cooperatives' services. The owners of a cooperative (who are also users) hold one or more types of stock in the cooperative. The fact that cooperatives do not distribute profits in proportion to ownership creates some unique structures and features of cooperative stock.

Member-Owner Control

The term *member* is used for an individual who has generally purchased one share of membership stock and otherwise met the requirements for membership. Members control the cooperative business. Most cooperative have a one-member-one-vote voting system. Some cooperative vote in proportion to business volume or use a combination of one-member-one-vote and business volume. The one-member-one-vote system leads to the description of such a system as being democratically controlled by its users. Members in a cooperative vote to elect board members as well as on important issues such as mergers, dissolution of the cooperative, or major structural changes. Members vote on any changes in the articles of incorporation of the cooperative, which make up its overall legal structure, and on changes to the bylaws, which describe the cooperative's membership, voting, and governance system. Members do not vote on day-to-day decisions, which are the responsibility of the manager, nor do they generally vote on policies. Policies are more specific than bylaws and cover everything from credit to the employee handbook. Policies are generally adopted and modified by the board of directors. The member role in a cooperative relates to voting and governance.

Member Benefit

The third unique feature of the cooperative business model is the fact that a cooperative distributes its profits to its users (customers) in proportion to business volume. We typically use the term *patron* to refer to a user (customer) who does business with a cooperative and is eligible to receive a share of profits in proportion to their use. The profits received are called a *patronage refund*. In most cases, a patron is a customer who is also a member of the cooperative. Cooperatives can do a portion of their business with nonmembers, so a cooperative may have both customers and patrons. As we will learn later, there are some types of cooperatives that distribute profits to both members and nonmembers. In any case, the patron role in a cooperative is one who receives profits in proportion to use.

The cooperative business model is also unique in that the cooperative members can receive benefits simply from the existence of the cooperative or the customer transaction. Many cooperatives were formed to provide services that were missing in the marketplace. Cooperatives were also formed to offset the power of large IOFs that controlled the market and were making excessive profits at the expense of their customers. For example, if a wheat producer had only one grain elevator in their region, they might find that the elevator offered unfavorable prices. The existence of the cooperative can therefore provide an invisible benefit in keeping the market competitive.

Focus on Member Needs

Cooperatives focus on meeting the needs of their member owners. They strive to provide the products and services needed by their members as efficiently as possible so that they can generate economic benefit (i.e., patronage refunds) to the members. In order to meet the long-run needs of members, they must retain a portion of profits to reinvest in the growth and stability of the cooperative. In evaluating new opportunities, cooperatives consider both their members' needs and potential profits. That creates a balancing act because the cooperative cannot provide economic benefit and ensure its long-term viability unless it is profitable. Still, it is often said that profits are not the only objective of cooperatives. Their true objective is to benefit their member-owners.

Agricultural cooperatives can provide benefits at both the farm and the cooperative levels. For example, sugar beets are a more profitable crop relative to the alternatives of corn or soybean. Producers can grow sugar beets only if there is a processing operation that can process the crop as it is harvested. Members of a sugar beet cooperative may therefore receive a benefit at the farm level by being able to grow a more profitable crop in addition to receiving their share of the cooperative's profit. For this reason, we sometimes consider agricultural cooperatives an extension of the farm business. This linkage between the cooperative and farm also creates challenges. Decisions regarding pricing, product lines, and services can be more controversial because the member-owners see them both in terms of customer (i.e., farm-level) effects as well as the impacts on the cooperative's profits and stability.

Differences Between a Cooperative and a Not-For-Profit Organization

While cooperatives focus on member needs, they are for-profit business organizations. Their goal is to generate economic benefit (which is usually in the form of profits) and return those profits to their user-owners. Note that this is different from a not-for-profit organization. In those organizations, the goal is to provide a service or pursue an objective. For example, the Oklahoma Agricultural Cooperative Council is a not-for-profit organization whose mission is to provide educational programs for cooperative leaders and influence legislation that might otherwise be adverse to cooperatives. If a not-for-profit organization has revenues in excess of expenses, it is prohibited by law from sharing those with its members. The funds can be used only to pursue the organization's objectives. When a not-for-profit organization is liquidated, the residual value must be given to another not-for-profit organization. As we will see, all those structures are different from those in a cooperative, where the objective is to create profits and share them with the member-users. When a cooperative is liquidated, the residual value is distributed to members in proportion to their past use of the cooperative.

The Various Roles in a Cooperative

There are a variety of potential roles or relationship with a cooperative. These are often described as *customer*, *patron*, *owner*, and *member*.

Customer

A *customer* is someone who uses a cooperative by buying products and services or selling commodities to it. A customer can have either a buying or a selling transaction. That is similar to a commercial bank, where a customer may be either a depositor or a borrower. Many cooperatives do a portion of their business with individuals who are not members.

Patron

Many people may use the term *patron* in a general sense that is interchangeable with *customer*. For example, a bank may refer to its depositors and borrowers as patrons rather than customers. When discussing cooperatives, we use the term *patron* in a narrower sense to describe a user who receives a share of the net income of the cooperative in proportion to their use in the form of a patronage refund. A patron is generally a member who also does business with the cooperative. The key difference between a customer and a patron is that the latter is a customer who receives a share of the profits.

Owner

An *owner* is an individual who makes an equity investment to help finance the assets of a cooperative. Because the benefits of a cooperative are distributed in proportion to use, almost all cooperative owners are also customers and patrons. Members can make equity investments directly by purchasing equity, which is often called *membership stock*. Members also make equity investments indirectly when they receive a portion of their profits in the form of equity that is called a *retained patronage refund*. Marketing cooperatives can also deduct a small equity investment for each unit of commodity marketed. This is called a *per-unit equity retain*. As we will see in the cooperative principles, one of the responsibilities of a cooperative member is to supply capital—or, in other words, to be an owner. Because the profit distribution is not related to the amount of ownership, cooperative members do not have an automatic desire to be owners and supply capital.

Member

A *member* is an individual who has a right to vote on the important affairs of the cooperative. In some cooperatives, a user must purchase a share of membership stock as part of the member application process. In that instance, they would be an owner—at least to a small extent—as well as a member. In other cooperatives, a user is allowed to purchase their membership equity share through retained patronage. In that case, they can be a member but not an owner. Among the important duties of a member are attending the annual meeting, voting on the election for the board of directors, approving of changes to the articles of incorporation or bylaws, and approving of mergers, major acquisitions, or the dissolution of the cooperative. The most common voting system in cooperatives is one member, one vote.

Pure Cooperative

A "pure" or ideal cooperative would be one in which all customers were members as well as patrons and owners. That would be idyllic because all the user decisions would be balanced across all perspectives. In an ideal cooperative, the individuals using the firm would also finance it in proportion to their use, receive an appropriate share of the profits, and be active in governing the cooperative. In practice, a cooperative may have some individuals who are not in all four roles. We have already mentioned nonmember businesses, where some customers are not members. A member could join and never do business with the cooperative and thus not be a patron or customer. A more typical situation is one in which users are not adequately represented in all four roles, which influences their behavior. For example, a new member may have made a very small investment and thus have only a nominal ownership role. An older member in a cooperative with a long equity revolving period may have a large investment but no longer be a user or a patron. User-owners who never attend an annual meeting or vote are not actually performing in their member role even if they are technically members. When the stakeholders in a cooperative are not participating in all four roles, the managers and directors face additional challenges.

Cooperative Principles

In the next chapter, we will cover the various versions of the cooperative principles. The cooperative principles that are used to guide today's cooperatives were first established in Rochdale, England, in the 1840s. The Rochdale

Society of Equitable Pioneers was created when a group of woolen mill workers formed a cooperative to purchase household supplies in volume. This group condensed centuries of cooperative experiences and formed a written set of cooperative principles. These principles were so carefully thought out that they are still in use today, some 160 years later! The Rochdale principles have been revised and modernized by various groups, including the U.S. Department of Agriculture and the International Cooperative Alliance, which created the set of principles that are most frequently referenced today.

The concept of a set of principles is unique to the cooperative business form. Cooperative principles are a set of ideals for the operation of cooperatives. They also identify the unique features of cooperatives that distinguish them from other business forms. Cooperative principles are often considered guidance as to whether a cooperative is structured and operating in a way that will make it a successful user-owned business over the long run. When we consider new structures and practices in the cooperative industry, we often consider whether they are consistent with the cooperative principles. As with any set of long-lasting principles, there is the eternal challenge of identifying which concepts are set in stone and which should be modernized in accordance with the current situation.

A Little Background on How a Cooperative Operates

Who Does What?

MEMBER

While the members have the ultimate control in a cooperative, they typically are not directly involved in most day-to-day decisions or even in setting policies or making long-term plans. If you or your family ever owned a mutual fund, you were probably at least an indirect owner in Walmart. That doesn't mean that you made the decisions on how your local store operated. Similarly, in a cooperative, members usually vote only on important issues, such as mergers, acquisitions, or dissolution of the cooperative. Members also vote to select the board of directors. Of course, members also have the responsibility of supporting the cooperative by doing business with it and staying informed on its status and activities.

BOARD MEMBER

The board of directors' roles are to oversee the cooperative's financial condition, set policies, and make long-range financial decisions. These might involve a taking out a long-term loan, retaining a portion of the profits as equity, or revolving previously issued equity for cash. It would not be efficient for members to vote to establish policies, and in many cases, policies might need to be changed before the next annual membership meeting. The board makes the long-run, or strategic, plan, although the manager often has input into the strategic planning process. The board is also responsible for authorizing the annual audit. Additionally, the board has a major role in selecting and hiring the manager, evaluating their performance, and setting their compensation. We often say that the board has one employee—the manager—and all other employees report to the manager. The board of directors has authority only when they are in a properly convened board meeting. At any other time, they have no more authority than any other member. Board members do have a communication role outside of board meetings, including listening to member input and communicating the board's actions and rationale while avoiding discussion of confidential aspects.

MANAGER

As mentioned, the manager oversees the employee group. The manager makes all the day-to-day or operational decisions. The manager makes short-run financial decisions on items such as budgets. They also provide financial and other information to the board so that they can perform their oversight role. The manager is involved with the annual audit, providing the auditor with all the information requested; however, the auditor provides the report to the board. In most cooperatives, the board policies establish a limit as to the expenditure level that the manager can spend without board approval.

Two Viewpoints on Cooperative Purpose

In this text, we will focus on the cooperative business model as a vehicle to create economic benefits and distribute those benefits to user-members. According to that viewpoint, the objective of the cooperative is to maximize the total profits at both the farm and the cooperative levels. Since cooperatives usually conduct transactions with their members at market prices, the maximum economic benefit is generally created by operating the cooperative as efficiently as possible and maximizing profits. Thus, the economic viewpoint on a cooperative is that it is a for-profit business that strives to create as much profit as possible and distribute it to its member-owners. In that economic view of the cooperative, we recognize the fact that the members are collectively doing what they could not do as efficiently on their own. We also realize that the existence of the cooperative may be keeping the marketplace competitive. However, we hope that we are operating the cooperative so that members gain economic benefits. In the answer to the question of why should one patronize a cooperative, economically, you will participate in reaping the profits and thus be better off.

There is another social view of cooperatives. According to this view, cooperatives are seen as a vehicle for creating economic opportunity and social justice. Some of those social goals may relate to keeping the marketplace competitive or generating market access; however, they may also go much further and include goals for improving the position of particular groups within society. Such social goals are laudable and may be supported by a wide spectrum of people. However, these social benefits can be achieved only if the cooperative is economically successful and remains in existence. Using any business organization as an engine for social change is dangerous because it creates conflicting objectives that might make it unable to survive in a capitalistic market economy. Another issue is whether the entire membership agrees on the social objectives and is willing to provide capital and forego profits to achieve them.

That being said, we recognize that social benefits and social justice are very important to individuals and society overall. There is a large body of literature emphasizing the social aspects of cooperatives. Some new forms of cooperatives, such as multistakeholder and worker cooperatives, have a strong focus on social goals. In this textbook, we will focus on the economic view of the cooperative organization. Students who are interested in the social benefits are encouraged to explore that dimension on their own.

Summary

Cooperatives are a unique form of for-profit businesses that are owned and controlled by its users and that distribute profits to those user-owners in proportion to their use. They operate under a unique set of principles and have exclusive structures for profit distribution and equity ownership. Most U.S. cooperatives operate under a traditional open-membership model, but the cooperative business model has also evolved to include newer forms, such as closed-membership cooperatives, hybrid member-investor cooperatives, and multistakeholder cooperatives. There are various roles of participation in a cooperative, including customer, patron, member, and owner. In addition to beginning your path to understanding what cooperatives are, you should also have an understanding of what they are *not*. Cooperatives are neither nonprofit corporations nor investor-owned firms such as partnerships, S or C corporations, or limited-liability companies. As you learn more about cooperatives, you will see how the structures for profit distribution, ownership, and governance in a cooperative contrasts with the systems used by noncooperative firms and organizations.

CHAPTER 2

Cooperative Principles

Introduction and Learning Objectives

This chapter introduces the reader to the concept of cooperative principles. These are unique to the cooperative business and help distinguish it from other forms of business. Cooperative principles originated from observations on what made a user-owned business successful. We can refer to the principles to ponder the question of whether we are operating like a true cooperative. While it is common to refer only to the most recent version of the principles, it is useful to review their evolution. Over time, the principles have tended to become less specific, and some of the historic values are now just considered to be business practices that each cooperative can evaluate relative to their situation. After reading this chapter, you should have a basic understanding of cooperative principles, their origin, and how they have evolved over time.

Unique Principles for a Unique Form of Business

Today's economy consists of millions of businesses that conduct the activities of production, distribution, exchange, and service functions. The concept of the business firm or business organization actually came out of an evolutionary process. In early societies, individuals and families produced only for their own subsistence. That gradually evolved to where families produced in excess of their own needs for the purpose of exchange with others. As society evolved—and particularly after the Industrial Revolution—large-scale production became possible, which created the potential for investors to pool their resources in funding business organizations.

As we are accustomed to seeing and dealing with business firms, we seldom consider their basic principles. A *principle* is a governing code of conduct or a fundamental theory or assumptions. It is difficult to find a list of the basic principles of investor-owned firms (IOFs) since they are taken for granted. If one were to develop such a list, it would probably state that the objective was to maximize profits, the investors who funded the firm should receive profits in proportion to their investment, the manager works in the best interest of the owners, and the firm sells to customers at prices that cover expenses and create a profit. Those principles and our concept of how a business should operate likely developed over time in response to structures that worked and those that didn't. For example, individuals who tried to get investors to provide capital without promising them shares of profit in proportion to their investment probably discovered that no one was willing to invest.

Cooperatives have several unique features relative to other forms of business. In particular, they have unique relationships between the customers and the business and the owners and the business. These unique relationships were discussed in Chapter 1. As you recall, customers can also be (1) patrons who receive a share of the profits based on business use, (2) owners who supply the capital, and (3) members who have voting power to govern the business. As opposed to the framework of IOFs, which evolved as mass production and widespread commerce became possible, cooperatives are a different system of organizing a business. Over time, many groups have experimented with different systems of organizing production and business activities, most of which did not prove to be sustainable. As a successful structure for user-owned businesses emerged, the groups organizing the new business form (i.e., the cooperative business model) documented the organization's principles.

While the definitions in the previous chapter described what a cooperative is, *cooperative principles* are the guidelines for how a cooperative should operate. They also distinguish cooperatives from other business forms. As you might imagine, there is no exact agreement on the guidelines for how a cooperative should operate, and the principles have evolved over time. Studying the cooperative principles helps provide a better understanding of the unique nature of cooperatives. Examining how the principles evolved also made it easier to identify the key principles and understand how we arrived at the most recent set of principles.

The discussion of principles in this chapter refers specifically to cooperative principles. Cooperatives, like any business, must adhere to good business principles. Principles such as maintaining integrity, creating value for customers, and monitoring and accountability apply to both cooperatives and IOFs. We will address some of those principles in the chapters on finance, marketing, and management. In this chapter, we will assume that the cooperative is being managed using good business principles and will concentrate on the unique principles of the cooperative business form.

Cooperative principles are just one factor in cooperative success. In order for a cooperative business to be successful, it must be effectively managed and financially successful. The cooperative business must also develop policies that provide more specific guidance as to how the firm should operate. Finally, the cooperative's daily practices must be consistent with its policies and principles. Some of the disagreement concerning cooperative principles revolves around whether a guideline or an action is a principle, policy, or practice. A definition of those concepts will help facilitate our discussion.

Principle

The term *principle* is defined as a governing code of conduct, general or fundamental truth, or comprehensive or fundamental law. In this case, the term *law* does not refer to actual legislation, although we will see in later chapters that elements of cooperative principles are reflected in cooperatives, enabling legislation. We think of principles as defining the essential characteristics and objectives of the cooperative business form.

Policy

A *policy* is a wise or expedient rule of conduct or management. It is not an unchanging truth but rather a highly recommended course of action based on the given situation. In the hierarchy of cooperative documents, we will discuss the articles of incorporation, bylaws, and policy manual. This definition of the word *policy* incorporates all the policies formally adopted by the cooperative's board of directors. Some elements of the bylaws could actually be policies in this sense of the word. The point is that while policies should be compatible and reinforce cooperative principles, policies are adopted by each cooperative based on their situation. For example, while "member democratic control" is a cooperative principle, the limiting of board members to two terms is a policy. Term limits are controversial. One cooperative might conclude that these limits helped get more members involved in governance, whereas another cooperative might decide that members can create their own term limits by voting out an incumbent director.

Practice

A *practice* is a usual method, customary habit, convention, or frequent and usual action. A cooperative's practices have to achieve its business objectives while also remaining compatible with its policies and cooperative principles. Some practices reflect the customary action but are open for changes as conditions dictate. For example, the cooperative may have a practice of having a meal and speaker featured at the annual meeting. Most cooperatives would not convert that practice to a formal policy because they realize that the cooperative might want to reevaluate the annual meeting each year. Some concepts that were once considered cooperative principles are now considered to be just practices. For example, the original cooperative principles included "cash trading only" as a principle. Most cooperatives now offer credit or accept credit cards, and cooperative leaders consider offering credit to be a practice that each cooperative can evaluate. The true test of whether a cooperative is complying with cooperative principles is whether its actual practices are compatible with them.

The Rochdale Cooperative Principles

It is generally agreed that current-day principles evolved from rules of conduct developed by the Rochdale Society in 1860. This cooperative operated for the first eight years under the Friendly Societies, and in 1852, it incorporated. A good review of the evolution of the Rochdale Society and the cooperative principles is provided in a paper published by the University of Saskatchewan Center for Cooperatives (Fairbairn, 1959).

The 28 founders of the Rochdale Society, often called the *Rochdale pioneers*, were from a variety of professions. The group consisted of nine weavers: two each of wool sorters and shoemakers and 15 in other trades, many of which were related to the weaving industry. They formed a consumer cooperative selling primarily consumer goods such as food and clothing because of dissatisfaction with the retail shopkeepers in their community. About half (15) were believers in socialism, and the remainder supported capitalism. They relied heavily on the experience of earlier cooperatives in putting together their own set of guidelines. The Rochdale pioneers were more realists than idealists. They wanted a cooperative business that would meet their needs and survive as a business. Using the experience of others as well as their own, they formulated and refined rules of conduct and points of organization for guiding the society's business affairs. The Rochdale pioneers did not propose a set of principles. However, many years later, their rules were studied by others and came to be called the *Rochdale principles*, or cooperative efforts in putting together their own guidelines.

Most authors list 12 Rochdale principles; however, others have slightly different lists (Barton, 1989). The rules described by the Rochdale Society evolved from its founding in 1844 up until the time they were published, in 1860. The original rules of conduct as published in the pioneers' annual almanac were as follows:

- Capital should be of their own providing and bear a fixed rate of interest.
- Only the purest provisions procurable should be supplied to members.
- Full weight and measure should be given.
- Market prices should be charged, and no credit should be given or asked.
- Profits should be divided *pro rata* upon the amount of purchases made by each member.
- The principle of "one member, one vote" should obtain in government and the equality of the sexes in membership.
- The management should be in the hands of officers and committees elected periodically.
- A definite percentage of profits should be allotted to education.

- Frequent statements and balance sheets should be presented to members.

As you can see, some of the first rules of conduct, such as "full weight and measure," "only the purest provisions," and "frequent statements and balance sheets," were likely recommendations of good business practices and not unique to the cooperative form. As mentioned, the Rochdale rules continued to evolve prior to being published in 1860. Rather than repeat the original text of the 1860 Rochdale principles, we present the principles here in current language and terminology while trying to retain their original meaning (Barton, 1989).

Rochdale Cooperative Principles

1. Net income distributed in proportion to patronage
2. Limited dividends on equity capital
3. Goods sold at market prices
4. One member, one vote
5. Open membership
6. Equity provided by patrons
7. Limited individual patron ownership
8. Duty to educate
9. Cash trading only
10. No unusual risk assumption
11. Religious neutrality
12. Equality of sexes in membership

There are various opinions as to how to treat the Rochdale principles. Some see them as the center of cooperative structure and philosophy and consider them timeless and of universal validity. Others claim that some of them were based on the unique situation facing the Rochdale pioneers. It is true that the conditions in Rochdale, England, in the 1840s are different from those in the United States today. More recent versions of cooperative principles have attempted to focus on the most universal principles and adapt them to current circumstances. (Reynolds, 2014) It is interesting to note that not everyone even acknowledges the evolution of the principles. Some most recent list of cooperative principles erroneously lists the most recent International Cooperative Alliance (ICA) principles as the Rochdale principles. Many authors apparently assume the ICA principles were simply restating the Rochdale principles without noting which principles had been dropped and which new concepts had been added.

Within the Rochdale principles, one can identify some that clearly relate to the cooperative model. The first, second, and sixth principles relate to the financial structure of cooperatives. Most observers consider those principles to be fundamental and universal, and we can see reference to them (albeit sometimes simplified) in every version of the cooperative principles. The reason for limiting dividends is not immediately apparent; however, if a cooperative paid out a high proportion of profits as dividends on invested capital, it would not have funds left for patronage distribution. In order to be primarily distributing profits in proportion to use, the cooperative must limit the distribution based on investment.

The concept of distributing benefits to users based on patronage is sometimes referred to as *operation at cost*. While that term is meant to imply that the goal of the cooperative to return profits to users, it should not be confused with the concept of a not-for-profit firm. As you can note from the third Rochdale principle, cooperatives typically exchange goods and services at market prices, thus generating a profit. The cooperative must retain a portion of the profits to finance new assets and revolve the older equity of members who are no longer using the cooperative. The term *operation at cost* is therefore a simplification and does not imply that cooperatives should not create accounting profits.

The concept of democratic control by user-members is also considered a fundamental principle, although not all agree that it should be so specific as to state as "One member, one vote." A broader interpretation is that members should have control in some structure that is consistent with the cooperative's user-oriented structure. Currently, about 35 states require one member, one vote in their basic cooperative legislation, whereas the remainder allow voting in proportion to patronage or a combination of the two. No state currently allows cooperative members to vote in proportion to ownership, although the hybrid member-investor statutes often allow the investor side to have ownership-based voting. As we will see, the most recent version of the cooperative principles (i.e., from the ICA) discuss democratic member control but specifically limit that to one member, one vote, presumably allowing voting in proportion to use.

Other Rochdale principles such as "Goods sold at market price," "Cash trading only," and "No unusual risk assumption" are now generally considered to no longer be fundamental principles but rather practices that each cooperative can consider relative to its circumstances. The concept of selling goods at market prices was intended to make sure the cooperative generated a normal profit, which could then be distributed. If a cooperative offers favorable prices—in essence, providing patronage up front—it runs the risk of underestimating costs and suffering a loss.

It is also interesting to note that the original Rochdale principles included some social goals such "Religious neutrality," "Equality of the sexes," and "Open membership." It should be noted that women's suffrage in England did not occur until 1928, so the Rochdale pioneers' goal of equality of the sexes was forward-thinking. While some may find the social goals within the Rochdale principles important, they have generally been dropped from later versions of the cooperative principles under the argument that they are not central to the cooperative business form. Some social goals reemerged in the most recent ICA principles.

Traditional Cooperative Principles

In the decades following the Civil War, the Grange and other organizations such as the American Farm Bureau Federation actively promoted farmer cooperatives. They adopted and modified the Rochdale principles into what are generally described as the eight traditional cooperative principles. Some of the traditional principles became part of state statutes for incorporating cooperatives. The traditional cooperative principles were referenced to guide cooperatives up until the 1980s (Barton, 1989). The traditional eight principles in concise form are as follows:

1. Voting is by members on a democratic basis (one member, one vote).
2. Membership is open.
3. Equity is provided by patrons.
4. Ownership of voting stock is limited.
5. Net income is distributed to patrons as patronage refunds on a cost basis.
6. Dividend on equity capital is limited.

7. Business is done primarily with member-patrons.
8. [There is a duty] to educate.

The traditional principles eliminated some of the Rochdale principles that related to business practices, such as cash trading and no unusual risk assumption. They also eliminated the social goal principles of equality of the sexes and religious neutrality.

By the 1980s, some of the traditional principles were considered to be outdated or simply good practices but not fundamental principles. The fourth principle, which related to limiting the ownership by any single patron, was considered unimportant if the first principle (member voting on a democratic basis) was in place, which separated control from ownership. The second principle (open membership) was considered impractical for cooperatives developing processing operations with a fixed capacity. However, there was still a strong perceived need to define the fundamental principles that distinguished cooperatives from other types of business organizations and to describe their obligations to users. What emerged is what is typically described as the contemporary cooperative principles.

Contemporary Cooperative Principles

The user-owner principle: The people who own and finance the cooperative are those who use the cooperative.

The user-control principle: The people who control the cooperative are those Who use the cooperative.

The user-benefits principle: The cooperative's sole purpose is to provide and distribute benefits to its users on the basis of their use.

If we examined the eight traditional principles and edited them to reflect what was eliminated (i.e., strike-outs) and what was modified (i.e., in italics), one could see the evolution of the principles. This comparison would retain the wording of the contemporary principles but illustrate what was deleted or added relative to the traditional principles.

Contemporary Principles Versus Traditional Principles

1. Voting is by members on a democratic (one member-one-vote) basis.
2. Membership is open.
3. Equity is provided by patrons.
4. Ownership of voting stock is limited.
5. Net income is distributed to patrons as patronage refunds on a cost basis (*based on their use*).
6. Dividend on equity capital is limited.
7. Business is done primarily with member-patrons.
8. [There is a duty] to educate.

While the wording is slightly different, one can see that those deletions and additions basically create the contemporary principles. The concept of "Business is done primarily with member-patrons" can really be combined with the seventh principle (net income distribution) to arrive at the contemporary third principle, which describes user benefits.

USDA Version of the Contemporary

In the mid-1980s, leaders of the cooperative industry had identified a need to update the cooperative principles. In 1987, the U.S. Department of Agriculture (USDA) helped prepare a report to U.S. Congress titled *Positioning Farmers' Cooperatives for the Future*. Although the report was directed at farmers' cooperatives, its principles were intended to apply to all types of cooperatives. The report represented the development of the contemporary principles. In later publications, the USDA often referred simply to the beginning statements of the principles, creating what is often referred to as the USDA's version of the contemporary principles (Reynolds, 2014).

USDA's Version of the Contemporary Principles

- User-owned
- User-controlled
- User benefits

The contemporary principles and the USDA's version of them can clearly be viewed as universal and timeless in the sense that they do not depend on social norms or economic conditions. Some view the contemporary principles as more of definitions of what a cooperative is rather than guidelines as to how it should operate. The contemporary principles clearly avoid describing any characteristic that could be considered by some to be simply a business practice. However, some in the cooperative industry, including the group who developed the ICA principles, felt that the contemporary principles were lacking in that they ignored the social and historical perspectives for defining the distinct purpose of cooperatives. In some senses, the contemporary and the USDA's versions of the contemporary principles were an economic view of the cooperative business model. They did a good job of distinguishing cooperatives from other business forms but did not meet everyone's goals for defining a unique cooperative identity.

The International Cooperative Alliance Principles

The ICA is a worldwide association of cooperatives and periodically approves a statement of principles (International Cooperative Alliance, 2014). A Statement on the Cooperative Identity was approved by the ICA at the 1995 General Assembly of the ICA in Manchester, England. The statement included a definition, description of values, and list and descriptions of seven principles. The statement was developed over a period of several years through consultation with its members. The ICA principles are very broad statements that address both economic and social interests. They have a much stronger social orientation than the four classes of principles reviewed previously. Control, ownership, and benefit guidelines are expressed in the first four principles.

ICA Cooperative Principles

1. Voluntary and open membership
2. Democratic member control
3. Member economic participation
4. Autonomy and independence
5. Education, training, and information
6. Cooperation among cooperatives
7. Concern for community

The first, second, and fourth principles provide control guidelines. The first ICA principle is similar to Rochdale principles five (open membership), 11 (religious neutrality), and 12 (equality of the sexes). The second ICA principle is similar to but less specific than the fourth Rochdale principle (one member, one vote). The third ICA principle provides guidelines for both ownership and benefits. It incorporates similar concepts to Rochdale principles one (patronage in proportion to use), two (limited dividends on capital), and six (equity provided by patrons). The fourth ICA principle likely related to government intervention in cooperatives in some parts of the world and has no similar counterparts in the other classes of principles. The fifth, sixth, and seventh ICA principles provide other types of guidelines. The fifth ICA principle reintroduces Rochdale principle 8, which was dropped in the traditional and contemporary versions. The sixth and seventh ICA principles have no similar counterparts in the other classes of principles. Those principles likely reflected the views of industry leaders regarding the unique aspects of cooperatives' identity.

The ICA principles clearly incorporate the basic three "defining principles" of user ownership, user control, and distribution of profits in proportion to use. As in previous versions of the principles, some cooperative leaders believe that additional principles should serve only as recommendations for business practices. Some believe that some ICA principles are not appropriate for all cooperatives in all business environments. For example, a housing cooperative could not have an open membership since it has a fixed number of housing units. Credit unions need to restrict membership based on financial criteria in order to avoid financial risks. Since the ICA principles are worded very broadly, one could also maintain that they don't restrict specific cooperatives from making those sort of sensible adjustments.

Summary

One of the distinctive features of the cooperative form of business is the promotion of and adherence to a set of principles. Other forms of business, such as IOFs, do not openly claim to abide by a set of principles. Principles play a central role in the cooperative culture and define to a great extent the nature and role of cooperatives. An understanding of the nature and role of cooperatives, including their definition, principles, policies, and practices, is of critical importance if you are to understand their nature, make them effective businesses, or compete with them in the most effective way.

Cooperative principles are embodied in the law and common practice. Some claim that they are also timeless and universally valid. However, significant disagreement exists over what constitutes the correct set of principles. Much of the controversy is caused by what we really mean by the terms *principle*, *policy*, and *practice*.

Cooperative principles have evolved over the past 150 years, beginning with the Rochdale principles. For purposes of discussion, we have identified four distinctive classes of principles: (1) Rochdale, (2) traditional, (3) contemporary, and (4) ICA. Current practice is most closely aligned with traditional and contemporary principles, although new cooperatives often endorse the ICA principles. Many believe that further evolution will or should occur if cooperatives are to continue as effective economic institutions.

Public policy has generally been favorable toward the creation and operation of cooperatives. State laws generally include special incorporation statutes for cooperatives, both agricultural and nonagricultural. Federal laws include the Capper-Volstead Act, which permits group action to form agricultural marketing cooperatives without being in violation of antitrust laws. They also include the special taxation provisions, which permit taxation of cooperative corporation income only once, either at the cooperative level or the patron level, if the income is distributed according to very restrictive rules. In general, net income distributed to users or patrons as patronage refunds is taxed once. All these laws involve legal definitions of cooperatives, and those definitions rely heavily on cooperative principles.

References

Barton, D. (1989). *Principles in cooperatives in agriculture* (D. Cobia, Ed.). Prentice Hall Publisher.

Fairbairn, B. (1959). *The meaning of Rochdale—the Rochdale pioneers and the cooperative principles* (Vol. Occasional paper series #94:02). Center for Cooperaitves University of Saskatchewen.

International Cooperative Alliance. (2014). *Guidence notes to the cooperative principles.* ICA.

Reynolds, B. J. (2014). *Comparing cooperative principles of the U.S. Department of Agriculture and the International Cooperative Alliance* (USDA Rural Development Rural Business Cooperative Programs Research Report No. 231). U.S. Department of Agriculture.

CHAPTER 3

Structure and Scope of Cooperatives

Introduction and Learning Objectives

When the word *cooperative* is mentioned, most people think of a farm supply cooperative, a dairy cooperative, or perhaps a natural food store. But the fact is that the cooperative business form is flexible and endlessly adaptable, and cooperatives provide almost every imaginable product and service a person could need from the cradle to the grave. Cooperatives exist in nearly every kind of business activity and are organized in a variety of ways. Like other types of businesses, cooperatives range in size from small business operations to massive and complex organizations involving thousands of members. Cooperatives can be classified into a few basic structural types in terms of membership and organization. But operationally, in terms of the products and services handled and functions performed, their relationships with members, other cooperatives, and other businesses are rich in variety.

This chapter discusses the function, structure, and scope of cooperatives with an emphasis on agricultural cooperatives. The objectives of this chapter are to help the reader: (1) develop an understanding of the variety of cooperative businesses and the scope of cooperative business activity in the United States and (2) become knowledgeable about the various cooperative organizational structures and strategic alliances.

Cooperative Structure

The structure of cooperatives may be viewed from many angles. We can classify cooperatives by their function, financial structure, member structure, layers of ownership, and type of members. Each is discussed here. Although these structures are not unique to agricultural cooperatives, they will be discussed within the agricultural context. When discussing cooperative operations and issues, it is common to mention the categories we discuss here. For example, in later chapters, we will describe how profits flow from federated regional cooperatives to local cooperatives. By having a basic understanding of the various dimensions of classifying cooperatives, you will be able to better understand the explanations and examples in later chapters.

Functions Performed

Cooperatives may be classified by the functions they perform or the commodities they handle. A cooperative's major activity is used to determine its classification by type, even though it may engage in multiple functions. Four general classifications are production, marketing, supply, and service. These groups may be further designated according to subfunction. For example, marketing cooperatives may bargain, process or manufacture, and sell products.

Production Cooperatives

Cooperative farming is found in some countries but rare in the United States; however, it is more prevalent in other countries. In the United States, a few dairy, hog, fruit, and vegetable cooperatives exist. In those cases, farmers have banded together to organize large-scale operations to achieve greater profits and value added. An example of a U.S. production cooperative is Ladder Creek Dairy, which was created when a group of corn and alfalfa growers pooled their resources to create a large-scale dairy cooperative. The dairy is, in essence, a processing operation to turn their corn and alfalfa into the value-added product of milk. A few farmer production cooperatives were organized in North Dakota from 1932 to 1981, when that state's Anti-Corporate Farming law was in effect. That law allowed individuals to organize their farms as cooperative corporations. A few farm cooperatives have been created in California and other states. Recently, there has been increased interest in production cooperatives in the Midwest. Many of those started as machinery-sharing cooperatives. As the members became comfortable with sharing machinery and sometimes operating labor across operations, some evolved to operate the combined acreage as one operation through a production cooperative.

Marketing Cooperatives

As their name implies, the primary function of marketing cooperatives is to market the products of their members. Marketing cooperatives are almost always also producer-owned or farmer-owned cooperatives. Marketing cooperatives cover a wide spectrum of activities but often focus on a single commodity or small range of commodities. Bargaining cooperatives negotiate with processors and other buyers and may never take title to or handle the commodity. Bargaining cooperatives are most prevalent in the fruit and vegetable sectors, in which the commodities are highly perishable and prices are volatile. At the other end of the spectrum, some marketing cooperatives also grade, process, package, label, store, and distribute products.

At times, multiple cooperatives pool their marketing efforts and create a separate cooperative called a *marketing agency*, which they share. Grain alliances that market grain for multiple cooperatives are a common example of marketing agencies.

The USDA no longer reports market share for cooperatives. According to the last published information, cooperatives marketed about 30% of all farm products. Cooperatives are particularly prevalent in the dairy sector, where they market more than 85% of the total milk supply. Cooperatives also have substantial market share in cotton, sugar, and rice. Cooperatives also market nearly every type of fruit, vegetable, and nut grown in the United States. They play a major role in marketing oranges, grapes, apples, cranberries, potatoes, and almonds. Many fruit and vegetable cooperatives market products under their own brands. You may be familiar with brands such as Blue Diamond, Ocean Spray, Sunkist, Sun-Maid, Sunsweet, Tree Top, and Welch's.

In recent years, new cooperatives have proliferated in the areas of organic and locally grown foods. Those firms run the spectrum from large cooperatives such as Organic Valley to single-location local food hubs. There are also numerous example of farmers' markets, which are organized as cooperatives.

Marketing cooperatives also play an important role in the livestock and poultry sector. These cooperatives are involved in selling and buying on commission and operating buying stations with a central desk. Some marketing cooperatives go beyond marketing and are involved in feeding, contract hog production, slaughtering and processing, and meat distribution. Approximately 14% of livestock in the United States is marketed through a cooperative.

Bargaining Cooperatives

Bargaining cooperatives, sometimes called *bargaining associations*, bargain or negotiate with processors and other first handlers for better terms of trade for members. Two or more operating cooperatives that are involved in the physical handling and/or processing of products may form a cooperative bargaining federation, often referred to as a *marketing agency in common*, to perform the bargaining function. Many dairy cooperatives combine bargaining with other operating and marketing functions. A pure bargaining cooperative does not physically handle or take title to the product involved but merely bargains for price and other terms of trade.

Processing or Manufacturing Cooperatives

The main emphasis of processing cooperatives is on the processing of raw farm products rather than on bargaining or marketing. For example, several dairy cooperatives that manufacture cheese, butter, and powdered milk leave marketing to brokers or a regional cooperative such as Land O'Lakes, Inc. Other examples include sugar and vegetable processing cooperatives. Some processing cooperatives operate in nonfood sectors, such as ethanol and soybean processing cooperatives.

Supply or Purchasing Cooperatives

Supply cooperatives purchase items in bulk and pass on the savings through profits and patronage. The term *supply cooperative* is typically used to describe a farmer-owned purchasing cooperative, whereas the term *purchasing cooperative* is used to describe a consumer-owned cooperative with similar objectives. Many farmers belong to farm supply cooperatives for their farm inputs, such as feed, seed, fertilizer, petroleum products, and hardware and building supplies. Supply cooperatives also provide farmers with crop protection products, such as insecticides, fungicides, herbicides, and soil treatments. Most farm supply cooperatives focus on bulk commodities, although some, particularly in areas with growing rural populations, may also operate retail operations. Some farm supply cooperatives also operate groceries, convenience stores, restaurants, and tire shops. In many cases, those operations evolved because no other business was willing to support the operations in a small rural community. Cooperatives supply more than 25% of all farm inputs and have a particularly prominent market share in bulk inputs such as fertilizer and petroleum products.

There are two categories of nonfarm-related purchasing cooperatives. There are retail and consumer purchasing cooperatives (which will be discussed in a subsequent section) and purchasing cooperatives that supply independent retailers. For example, the owners of independent hardware stores organized purchasing cooperatives to achieve economies of scale in purchasing. That helped them compete with big warehouse chains such as Home Depot and Lowe's. More than 6,000 True Value dealer-owned hardware stores are members of the TruServ purchasing cooperative, which was established in 1948. The Ace Hardware dealer-owned cooperative was established in 1924 and now has 4,800 dealer members. Similarly, fast-food restaurants have formed purchasing cooperatives. Those cooperatives, which are owned by the franchise store owners supply food, restaurant supplies, equipment, advertising, insurance, and almost anything the fast-food operation needs. Restaurant franchises such as Burger King, Dairy Queen, Kentucky Fried Chicken, and Taco Bell have all organized purchasing cooperatives. That structure creates confusion because the retail businesses (e.g., an Ace Hardware store or a Burger King restaurant) are investor-owned firms (IOFs) organized as limited liability companies or corporations. However, the business owners or the business entities are members of a purchasing cooperative.

In several parts of the country, hospitals have formed purchasing cooperatives to maximize their purchasing power on expenditures such as laboratory products, food, film, pharmaceuticals, and other services. Purchasing cooperatives have also been formed by independent neighborhood pharmacies.

Service Cooperatives

The functional category of *service cooperative* is commonly used to describe a subset of agricultural or farmer-owned cooperatives, although there are consumer-owned cooperatives, such as day care cooperatives, that provide services. Agricultural service cooperatives provide a wide range of services, including artificial insemination, milk testing, trucking, crop drying, and livestock shipping. Because cotton gins do not purchase the members' cotton but rather gin the cotton (i.e., separate seed from lint) and deliver it to farmer-owned warehouse cooperatives, cotton ginning and cotton warehouse cooperatives are sometimes considered service rather than marketing cooperatives.

Finance, utility, and insurance cooperatives are also sometimes classified as service cooperatives. Important financial cooperatives within the agricultural sector include the Farm Credit System (FCS). Established in 1916, the FCS is the oldest and largest financial cooperative in the United States. Organized as cooperatives and supervised by an independent government agency, the Farm Credit Administration includes CoBank, which lends to cooperatives, and federal land banks and Production Credit Associations. The FCS provides loans, crop insurance, and other services to more than a half-million farmers, agribusinesses, and cooperatives. It is estimate that the FCS provides more than 25% of agricultural credit in the United States.

CoBank, the national cooperative bank providing credit to cooperatives, is part of the FCS. It was created in 1989 as a result of mergers between the original Bank for Cooperatives. Other financial cooperatives that serve cooperatives (not necessarily agricultural) include the National Rural Utilities Cooperative Finance Corporation, which has provided funds to rural electric and telephone cooperatives since 1969. The National Cooperative Bank is a leader in providing loans to housing, consumer, and other nonagricultural cooperatives in the United States.

In order to be eligible to obtain a loan from CoBank, a business must:

1. Have 80% of its voting control in the hands of agricultural or aquatic producers or federations of cooperatives, with voting control being as low as 60% for certain supply farm cooperatives
2. Do a minimum of 50% of its business with members, except that which is done with the government and services and supplies furnished by the cooperative as a public utility, which are excluded
3. Use one-member-one-vote governance or limit the capital stock dividend rate to the lesser of the minimum allowed under state law or 8%

As we discuss new cooperative structures and what distinguishes a cooperative, one can see how the CoBank eligibility comes into play. New cooperative structures or hybrid member-investor structures are likely not eligible to borrow from CoBank.

Credit Unions

Credit unions are the most prevalent consumer-owned financial service cooperative. As of 2022, there were more than 4,700 credit unions, with almost 132 million members. In terms of membership, credit unions are the largest membership in the United States, and their membership represents 45% of the economically active population in the United States. Credit unions were first established in the early 1900s to provide financial services to people with a common bond—employees of particular companies, members of churches or unions, and people who lived in the same neighborhood. Today, we can find credit unions for schools and universities, federal employees, communities, companies, and elsewhere. Increasingly, low-income communities have come to view credit unions as a force for economic development—that is, a means of recycling neighborhood money for neighborhood purposes. In addition to auto loans and home mortgage loans, they also offer many other services, such as credit cards, automatic teller machines, deferred retirement accounts, certificates of deposits, and even mutual funds. Credit union cooperatives have a unique structure relative to other financial cooperatives in that they both source and lend funds from their members. Credit union depositors create the funds that are used by credit union borrowers. Other financial services cooperatives such as CoBank or FCS banks acquire funds by selling bonds in the financial marketplace.

Utility Cooperatives

Another category of service cooperatives are rural electric cooperatives (RECs) and telephone cooperatives. RECs provide electricity to about 56% of U.S. land mass. They own and maintain 42% of the electrical lines in the United States while serving only 12% of the population. RECs average 7.4 customers per mile as compared with investor-owned utilities (IOUs), which average 34 customers per mile. It was because of the low customer density that IOUs chose not to serve rural areas. RECs can be found in 2,600 of the more than 3,100 counties in the United States, Puerto Rico, and American Samoa. Currently, there are more than 900 RECs, with 19 million consumer members and revenues more than $45 billion. They primarily obtain their power from 62 generation and transmission cooperatives that they own.

RECs, which are consumer-owned, locally managed entities, proliferated during and after the Great Depression, when the vast majority of U.S. farmers didn't have electricity. Traditional utilities had refused to provide service to rural America because they claimed it wasn't cost-effective. By the mid-1930s, only one in 10 farms had electric power, and rates were far higher than those paid by people living in town. The potential of extending electric power to the countryside was widely recognized by government leaders and particularly farmers, even though IOF power companies expressed little interest in expanding their service. The Rural Electrification Act (REA) of 1936 set up the Rural Electrification Administration in the U.S. Department of Agriculture (USDA) as a lending agency to finance the extension of electric power to rural areas. Farmers moved quickly through cooperatives to take advantage of the new program. As a result, a formidable argument could be advanced that RECs are responsible for bringing about one of the more profound changes in U.S. agriculture. While the REA provided initial loan funding, RECs repaid all of it, achieving a rural infrastructure of electrical power funded entirely by user-members.

On October 28, 1949, the REA was authorized to receive the loan money necessary to improve and extend telephone service in rural areas. The Rural Telephone Bank was established within the USDA in 1971 as a source for financing rural telephone companies. There are currently 260 telephone cooperatives providing telecommunication services to 1.2 million rural Americans in 31 states. In addition to telephone cooperatives, there are close to 3,000 water cooperatives with 2 million members in the United States. Like RECs and telephone cooperatives, water cooperatives were created in areas where the population density was too low to attract investor-owned projects.

Other Service Cooperatives

The Associated Press (AP) is a service cooperative organized in 1848 by six newspapers in New York City. The purpose was to join forces and bear the high cost of the latest news-transmitting technology. Since then, the AP has become the world's largest and most important news service, serving 98.8% of all the daily newspapers in the United States and about 6,000 radio and television stations.

These few examples demonstrate the adaptability of cooperative structure to most any situation. Additionally, farmers have extended the scope of cooperatives through organizational combinations.

Financial Structure

Cooperatives are incorporated as either capital stock or noncapital stock organizations. The type of capital structure is specified in the articles of incorporation.

STOCK COOPERATIVES

Most agricultural cooperative are stock cooperatives. In a stock cooperative, a member purchases or earns a share of membership stock, which is associated with voting rights. In open-membership stock cooperatives, members acquire additional stock or capital credits through retained patronage. In a defined membership stock, cooperatives' members purchase additional shares of stock with associated usage rights. Some stock cooperatives issue preferred stock that may owned by both members and nonmembers. Preferred stock often pays a dividend but does not have voting rights. In contrast with IOFs, which have general provisions for the ready transfer of stock ownership, stock cooperatives usually have restrictions on the transferability of stock by members. The purpose of the stock transfer restriction is to limit the ownership of the cooperative primarily to qualifying member patrons.

NONSTOCK COOPERATIVES

Nonstock cooperatives emerged out of the historical development of cooperatives in the United States. Before the Capper-Volstead Act (CVA) was passed in 1922, which led to the current structure of agricultural cooperatives, various groups advocated forming cooperatives, and states attempted to modify corporate law to allow the cooperative form of business. Critics of these initial cooperative efforts felt that because the cooperatives issued stock, they could become just another form of an IOF. They were also concerned that the structures and legislation did not prohibit the stock from being sold to a nonmember. Some agricultural groups sought to remedy those perceived problems by creating a nonstock cooperative model, which they felt was a pure cooperative. Legislation for nonstock cooperatives emerged in several states. Some of the laws sought to avoid capital stock by putting all member capital on a loan basis.

The statues also emphasized a strict operation at cost where goods would not be priced at market prices but rather at costs with a fee to cover operating expenses. In a sense, the nonstock cooperative was perceived as purely an extension of the farm firm. Nonstock laws were developed in California, Alabama, and Texas. A few cooperatives that were incorporated as nonstock cooperatives in those states still operate successfully. The nonstock cooperative received a brief revitalization after the passage of the Clayton Act in 1914, which provided nonstock cooperatives with immunity from antitrust prosecution. That became unimportant when the CVA was passed in 1922, which, as we shall learn in later chapters, created the structure for U.S. agricultural cooperatives and provided both stock and nonstock cooperatives with antitrust immunity.

Most existing nonstock cooperatives operate very similar to stock cooperatives, using membership certificates and/or capital certificates in the place of stock. Some classes of the capital certificates are typically issued as part of patronage refunds, creating revolving equity. Interestingly enough, nonstock cooperatives do not need a separate structure to offer equity to nonmembers since the capital certificates have neither voting rights nor restriction on ownership. Just as preferred stock is a very small portion of cooperative equity, very little capital certificates are owned by nonmembers.

The concept of replacing member equity with loans in a nonstock cooperative was generally unsuccessful since it suggested am obligation to repay the capital at a predetermined rate. Owners in any firm should receive the residual profits, which is consistent with the concept of revolving equity based on the financial condition of the cooperative. The strict concept of operating at cost was also problematic because the cooperatives often could not predict costs or faced unanticipated risks. Because they had in essence distributed the anticipated profits up front through pricing, they experienced losses and financial instability.

While nonstock cooperatives are mostly a historical holdover, they do occasionally create interesting issues. For example, because cooperative laws are very brief, cooperatives often fall under general corporate laws for issues such as board of director responsibility or stockholder rights. The absence of stock occasionally makes it difficult to determine what legal precedent applies to nonstock cooperatives. The early attempts at cooperative forms and legislation in the United States also helps us to understand why the CVA was needed at the federal level and to appreciate its structure.

Open Versus Defined Membership

While they are described as differences in membership structure, open-membership and defined-membership cooperatives also have fundamentally different financial structures. As discussed previously, open-membership cooperatives have a small membership investment and create most equity through retained patronage, which is revolved by the cooperative and not exchanged between members. Defined-membership cooperatives obtain most or all their equity through up-front membership investment, with each share of stock having an associated usage right. Equity in a defined-membership cooperative can be bought or sold to another qualifying individual with the approval of the board, creating the potential for stock appreciation. While similar in most other respects, the equity structure of open and defined cooperatives is quite different.

LAYERS OF OWNERSHIP

Most agricultural cooperatives in the United States were started by groups of farmers within a rural community. That created the concept of what we refer to as a *local cooperative*. Initially, local cooperatives were independent and operated in relatively small geographic areas, typically within a radius of 10 to 30 miles or within a single county. They were managed out of a single office. At the same time, some cooperatives were organized in hopes of operating over a large geographic area, such as an entire state. That created the concept of a regional cooperative. Mergers and acquisitions have enlarged the operating size of many local cooperatives, some of which have now grown to the size of what once would have been considered a regional cooperative. While the geographic classification of cooperatives has become less relevant, the layered ownership structure of some regional cooperatives is important to understand,

Centralized cooperatives (regardless of whether they are local or regional) operate under a single manager and home office. The members are direct members of the cooperative. There is a single board of directors, although the board seats may be assigned to districts. Centralized cooperatives can operate efficiently since operational and logistic decisions are coordinated. As a centralized cooperative gets larger, member communication becomes more difficult since the manager and board can be located far away from many local members.

A federated cooperative is a cooperative owned by other cooperatives. For that reason, almost all federated cooperatives are regional, whereas not all regional cooperatives are federated. In the federated structure, producers are a member of their local cooperative, which is a member of the region. Producers serve on the board of the local cooperative, while a representative, often either a local board member or manager, serves on the board of the regional cooperative. That structure improves communication with the local producer. The federated system gives much more power at the local level. Just as how a producer does not have to do business with their local cooperative, local cooperatives to not have to do business with their federated regional. The cost of that flexibility is efficiency. A federated regional cooperative will find it more difficult to coordinate logistics or use facilities since each local members can decide whether to participate. It is important for potential members and/or employees to understand the differences between the structures so that they can distinguish between headquarters and a branch office.

NATIONAL AND INTERNATIONAL COOPERATIVES

Some regional cooperatives have expanded to the point that they operate through much of the United States and have significant international operations. It is easy to see why commodity marketing cooperative would want international operations to create additional marketing opportunities. It is also advantageous for processing cooperatives to have international operations since they can source from multiple hemispheres and achieve a year-round supply. Around the world, there are many cooperatives operating in multiple nations, and some European dairy and genetics cooperatives operate in the United States. There is really no structural distinction between a regional, national, or international cooperative; however, the terms *national* and *international* emerged as it became evident that some regional cooperatives operated in an area much larger than a region.

COMBINATION OF CENTRALIZED AND FEDERATED COOPERATIVES

Many large regional cooperatives have features of both federated and centralized structure. In these cooperatives, both individuals and autonomous cooperatives are direct members. For example, in Land O'Lakes, dairy producers are direct members, while the local cooperatives are members on the farm supply side. CHS has both member local cooperatives and also operates a number of elevators and farm supply operations, which are directly managed out of the headquarters. Both individual growers of Sunkist and a local Sunkist packinghouse are direct members. Southern States has mostly the structure of a centralized cooperative with direct farmer members but also some affiliated cooperative members, making it technically also a combination structure.

The major problem with the combination type of structure is the difficulty in determining voting rights. One solution has been to allow one vote to each member cooperative and one to each individual, plus additional votes based on the volume of business that each local does with the regional.

Other Structural Arrangements

Cooperatives can use several other types of structural arrangements to take advantage of economic opportunities. The objectives of such arrangements are to gain efficiencies in operation, enter into other activities, enhance financial strength, gain market entry, increase market power, capitalize on core competencies, and reduce competition among cooperatives.

A ***subsidiary*** corporation is a corporation organized, owned, and controlled either directly or through trustees by a parent cooperative. To legally isolate the parent cooperative from the subsidiary, there needs to be clear separation of management and profits between the two organizations. The purpose of the subsidiary is to assume certain duties and functions of the parent cooperative.

A ***marketing agencyincommon*** is organized by two or more marketing cooperatives to market the output of member cooperatives. A marketing agency in common allows local cooperatives to merge all of their marketing operations without assimilating the cooperatives. Much of the grain handled by Oklahoma cooperatives is marketed through a two-state marketing agency in common: CoMark Equity Alliance, headquartered in Enid, Oklahoma.

A ***joint venture*** is an association of two or more participants, persons, partnerships, corporations, or cooperatives to carry on a specific economic operation, enterprise, or venture. The identities of these participants, however, remain separate from their ownership or participation in the venture. Use of joint ventures among cooperatives involves a partnership arrangement between two or more cooperatives. This type of activity has been commonplace among regional cooperatives marketing cotton, grain, dairy products, and fruits and vegetables.

Regional purchasing (supply) cooperatives have formed joint ventures to manufacture feed and fertilizer or refine petroleum products. More recently, cooperatives have become involved in joint ventures with IOFs.

An ***information-sharing organization*** is two or more cooperatives that market and price independently but exchange production and market information. Improved production and market information allows the participating cooperatives to improve marketing and pricing of their products. Twelve dairy cooperatives that process and market about 60% of the dry whey and lactose in the United States exchange weekly their production, inventories, and market information

Classification by Ownership

Cooperatives can also be classified by the type of member by whom they are owned. The major categories are farmer-owned, consumer-owned, business-owned, and worker-owned. The differences in ownership affects the cooperative's objectives, structure, and sometimes taxation. Agricultural cooperatives operate as an extension of the farm firm providing economies of scale in purchasing inputs and marketing commodities. Farmer members tend to have substantial investment in their cooperatives, primarily through retained patronage. Farmer patrons are taxed on their patronage distributions.

Consumer cooperatives are primarily credit unions and retail operations. Consumer cooperatives often do not retain equity as revolving equity but instead simply retain it as unallocated retained earnings. Even if that retained patronage was allocated, it would make up a very small portion of the average consumer member's net worth, so the practice of retaining profits as unallocated equity is not controversial. RECs and other utility cooperatives operate under the system of revolving equity. However, they tend to have very long equity revolving periods, with some RECs redeeming the equity only in the event of the death of a member or having the estate receive it under the revolving cycle. One rationale is that the member's retained equity is helping to keep the utilities interest rates and interest expense lower. REC members therefore benefit from their retained patronage by keeping expenses and electricity rates lower.

Consumer cooperatives generally provide goods and services purchased with after-tax income. For that reason, members of consumer cooperatives are not taxed on patronage. Some retail consumer cooperatives, including Recreation Equipment, Inc., the largest consumer cooperative in the United States, issue patronage in the form of credits on future purchases. That practice would be difficult in an agricultural cooperative because of taxation issues. Other categories of consumer cooperatives include housing cooperatives, insurance cooperatives, and utility cooperatives.

Business cooperatives typically purchase cooperatives owned by IOFs, often the owners of franchise outlets. They face the same taxation structure as agricultural cooperatives since the cooperatives are in essence an extension of their business. Because purchasing cooperatives operate in the background providing savings to business owners, they tend to be an invisible aspect of the U.S. cooperative sector.

Worker-owned cooperatives, such as taxicab cooperatives, are owned by the employees. The worker-owned sector is the smallest cooperative sector in the United States. In a worker-owned cooperative, patronage is distributed in proportion to hours of work or some other measure of employee contribution. Some worker cooperatives have different objectives relative to other ownership structures. Most cooperatives seek to maximize efficiency and keep wage expenses as low as possible while retaining quality employees. The objective of worker cooperatives is to maximize the benefits to the employees through both wages (which are typically set at or below market rates) and patronage. Examples of worker-owned cooperatives include day care cooperatives, education cooperatives, and worker-owned home health care cooperatives.

Scope of Total Cooperative Activity

According to the most recent (2019) USDA agricultural statistics (USDA, 2019), there are 1,871 agricultural cooperatives in the United States, with $197 billion in sales and $6.2 billion in net income. Commodity marketing made up $119 billion in sales, while farm supply sales were $24 billion. Dairy marketing cooperatives had the highest sales volume ($39 billion), followed by grain marketing ($30 billion), livestock ($3.5 billion), and cotton ($2.7 billion). According to the USDA statistics, 1.89 million individuals are members of U.S. agricultural cooperatives.

The University of Wisconsin Center for Cooperatives (UWCC) conducted a research project on the scope of the entire cooperative industry in the United States and its economic impact (Deller, 2009). The UWCC study found that there were 30,000 cooperatives in the United States in 73,000 separate locations with $3 trillion in assets and more than $500 billion in revenue. Americans hold 350 million memberships in cooperatives, with nearly 340 million memberships in consumer cooperatives. Cooperatives generate nearly $79 billion in economic benefits from patronage refunds and redeemed equity.

Table 3.1 U.S. Cooperatives by Type

Summary of Key Economic Indicators							
Type	**Assets (in Million USD)**	**Revenue (in Million USD)**	**Wages (in Million USD)**	**Firms**	**Percentage of All Cooperatives**	**Employees (in Thousands)**	**Memberships (in Thousands)**
Worker	128	219.04	55.41	223	1%	2.38	55.14
Farmer	23,632	65,436	2,970	1,494	5%	72.93	714.65
Purchasing	1,126,848	157,092	2,902	724	2%	130.35	6,133
Consumer	1,975,805	291,086	19,085	19,844	92%	650.65	343.969
Total	**3,126,414**	**514,524**	**25,013**	**29,013**	**100%**	**856.31**	**350,872**
USD = U.S. dollar.							
One individual can belong to multiple cooperatives, so the total number of individuals who are members may be less than the total membership. Purchasing cooperative are primarily owned by businesses.							
Adapted from Steven Deller et al., "U.S. Cooperatives by Type: Summary of Key Economic Indicators," Research on the Economic Impact of Cooperatives. Copyright © 2009 by University of Wisconsin Center for Cooperatives.							

Note that the UWCC study was based on a survey and did not capture all cooperatives. You might note that the USDA statistics show a higher number of agricultural cooperatives with substantially higher membership and higher revenue. The UWCC's are the only data available for the entire cooperative sector in the United States and are very useful in illustrating the relative size of the subsectors. You might note that 92% of all cooperatives in the United States are consumer cooperatives. They also account for the vast amount of cooperative membership. Purchasing cooperatives (business supply cooperatives) have relatively few members but substantial sales. Approximately 5% of all cooperatives are agricultural cooperatives. They account for only 0.2% of all membership but 12.7% of revenue. That reflects the fact that the typical agricultural producer member does a much larger dollar business volume with their cooperative relative to a typical consumer member. Agricultural cooperatives are very important in the agricultural sector, but just as agricultural is a small portion of the U.S. economy, agricultural cooperatives represent a small portion of the entire cooperative sector.

Summary

Cooperatives are diverse, and there are examples of successful cooperatives in virtually all industries and business segments. There are many ways to classify cooperatives, including by function, membership structure, financial structure, ownership structure, and owner characteristic. Within the classification of ownership, there is the federated structure, which involves multiple layers of ownership. All these classifications provide a better understanding of the diversity of cooperative businesses. After reading this chapter, you should have a better understanding of the diversity within the cooperative business model. This understanding of structure and scope will also help you understand our subsequent discussion of cooperatives.

References

Deller, S. H.-S. (2009). *Research on the economic impact of cooperatives.* University of Wisconsin Center for Cooperatives.

United States Department of Agriculture. (2019). Agricultural cooperative statistics (USDA Rural Development Cooperative Service Service Report 83).

CHAPTER 4

Nontraditional Cooperatives

Introduction and Learning Objectives

The majority of agricultural cooperatives in the United States are organized as open-membership cooperatives. There are other structures of cooperatives that have different property rights characteristics. It is useful to understand those newer cooperative models and their advantages and disadvantages. After reading this chapter, you should have a basic understanding of the closed-membership, hybrid member-investor, and multistakeholder cooperatives.

Nontraditional Cooperatives

Our discussion so far has focused on a traditional open-membership cooperative in which members can join at any time with a small membership investment and most of the equity is created through the system of retaining patronage and revolving equity. Most of the agricultural cooperatives in the United States operate under the open-membership structure. The structure has worked well for commodity farm supply and marketing cooperatives. Under the open-membership structure, producers can join the cooperative with minimum investment and build ownership through retained patronage. Nonmembers have an incentive to join in order to qualify for patronage. The infrastructure investment and capital needs of open-membership cooperatives are moderate, relative to processing cooperative, so traditional cooperatives can fund growth through retained profits.

When producer interest expanded from supplying and marketing commodities to manufacturing value-added products, disadvantages with the open-membership model were revealed. Value-added processing required a large up-front investment and had a fixed capacity. Value-added activities often had the potential for high returns but also a high level of risk. Those characteristics did not match up with the open-membership financial model. Members were reluctant to make a large investment in membership stock that could not be bought or sold. There was no incentive to make the risky investment to start the cooperative since it was possible to join at a later time, after success was demonstrated. The defined-membership cooperative, often called the *new-generation cooperative* or *closed-membership cooperative*, was developed to address those issues.

Defined-Membership Cooperatives

Defined-membership cooperatives operate under the same basic legal structure as traditional cooperatives but have different financing and member commitments. While the member financing and marketing contracts differ, defined-membership cooperatives distribute profits in proportion to use and typically use a one-member-one-vote system. There are some distinguishing characteristics to the defined-membership form. Defined-membership cooperatives are structured around a fixed capacity, which is allocated to members through usage rights. Each share of stock has an associated usage right. For example, in an ethanol cooperative, a producer would have to own one share of stock to have the right to deliver one bushel of corn. Under that structure, patronage is linked to investment, so profits are simultaneously distributed in proportion to use and to investment.

The usage rights in defined-membership cooperatives are also tradeable. A member wishing to exit the cooperative can sell their usage rights to another eligible member with the approval of the board of directors. That structure makes the membership stock investment much more liquid and creates the opportunity for it to appreciate. If a newly created defined-membership cooperative is profitable, a member could have the opportunity to sell their usage rights at a substantial gain. In some defined-membership cooperatives, the usage right is also an obligation. That gives the cooperative a stable supply, keeping the firm at optimal capacity. Defined-membership cooperatives require a high up-front investment to join but often pay high portions of cash patronage since they obtain equity through the membership stock.

Hybrid Member-Investor Cooperative

Another nontraditional cooperative form is the hybrid form, which combines the user-member and outside investor capital. Many states have developed new statutes called *limited cooperative association statutes* to facilitate this alternative cooperative form, which is more conducive to outside investment. Hybrid member-investor cooperatives have two types of membership: user-members and outside investor-owners. The earnings of the cooperatives are allocated between the two classes under a pre-agreed-upon formula. The user-members receive benefits in proportion to use, and the investor-owners receive benefits in proportion to investment. The statutes place limitations on the maximum voting rights and profit distribution to the investor-members. As the name implies, the member-investor cooperative is actually the combination of a cooperative and an investor-owned firm.

The hybrid member-investor cooperative was developed as a vehicle to access outside capital. The defined-membership cooperative model improved incentives for producers to make up-front investment in more capital-intensive cooperative projects. The pool of investment was still limited to producers who were able to invest and deliver to the cooperative. Many proposed cooperatives were unable to raise sufficient equity. The goal of the hybrid member-investor model was to bring in funding from nonproducers while still maintaining a portion of profits and control to producer-members.

There are inherent challenges with the member-investor model. The two membership groups have different incentives. Issues such as commodity prices are controversial. If a marketing cooperative were to offer favorable prices, the user-members would benefit from the prices at the expense of the investor-members, who would receive less profit. Many new cooperatives incorporate under the member-investor statutes just to maintain flexibility. So far, there have not been many positive examples of merging investors with user-members.

The member-investor model also creates issues with compliance with cooperative principles. Cooperative principles suggest that members should vote on a one-member-one-vote basis or in proportion to the business model. Investor-members in a hybrid cooperative generally vote in proportion to investment, which is not in accordance with cooperative principles. The investor-members also share in profits in proportion to ownership, which differs from the cooperative principle of benefit in proportion to use. Hybrid member-investor cooperatives would likely not be eligible for Capper-Volstead Act protection and may face issues with CoBank eligibility, Subchapter T taxation, and U.S. Securities and Exchange Commission security registration exemption. All of those topics are discussed in more details in subsequent chapters.

Multistakeholder Cooperatives

Many traditional cooperatives have members who use different aspects of the cooperative. For example, a grain and farm supply cooperative may have producers marketing grain and other producers who only purchase inputs. Credit unions have both depositors and borrowers. In both examples, the members are still in a similar group (producer or consumer). In a multiple-stakeholder cooperative, there are more than one type of member. For example, a local food cooperative could have both consumer-members and producer-members as well as have worker-members. Some stakeholders discuss multistakeholder cooperatives as being part of a solidarity model. They want to expand the scope of the cooperatives' value package to a broader set of stakeholders.

The multistakeholder model requires defining multiple groups of membership and allocating governance and patronage across the groups. The multistakeholder model involves more individuals from the cooperative and provides incentives to additional groups. For example, bringing the employee group in as members provides them with skin in the game and additional incentives to remain as long-term employees. The multistakeholder model can also increase the pool of equity capital since additional groups invest or receive profits in the form of equity.

The disadvantage of the multistakeholder model is that it is more complex and creates greater potential for conflicts. The various stakeholder groups obviously have different incentives. For example, worker-members would lobby for increased wages, whereas producer-members would prefer reducing personnel costs. In designing their structure, the cooperative must decide the investment of each group, how patronage is split across the groups, and what the governance rights of each group are. The various groups often have different time horizons and economic situations. The structure may lead to dominance by one group or provide each group with equal power and benefits. Both alternatives can lead to difficulties. Even if the cooperative is able to navigate a fair structure, one group may evolve to dominate the cooperative over time. For example, there may be more members in one group, or one group may have better information and be more active in governance.

There are examples of successful multistakeholder cooperatives, including Weaver Street Market in Carrboro, North Carolina. Weaver Street Market is a worker- and consumer-owned cooperative selling natural, organic, and locally based foods and operating at multiple locations in North Carolina. The worker-members and consumer-members have equal representation on the board of directors. In recent years, many groups have examined the success of Weaver Street Market and investigated the multistakeholder model.

Given the challenges of navigating the multistakeholder model, many groups ask whether there are alternatives. There are partial alternatives within the single-stakeholder model. Consumer cooperatives often encourage their employees to become consumer-members. That does not align the employee interest with the consumer-members, but it does provide more employee engagement and understanding of the cooperative model. Multiple patronage pools facilitate different subclasses but do not actually address the question of additional types of members. Some cooperatives appoint a nonvoting employee delegate to report to the board of directors. That provides a partial alternative to create more employee engagement and loyalty.

These nontraditional cooperative examples illustrate the fact that the cooperative model is not static. In some cases, new forms of cooperatives are structured under existing statues, and, in some cases, new statues are developed. These innovative structures raise issues as to whether the entities are true cooperatives, compliant with cooperative principles. These nontraditional structures also make it difficult to make definitive, all-inclusive statements about cooperative equity, profit distribution, and governance.

Summary

The most appropriate form of cooperative is the form that fits the situation. Closed-membership, hybrid member-investor, and multistakeholder cooperatives have different structures. These structures address some limitations of the traditional open-membership structure but have their own limitations. The nontraditional structures prove that the cooperative business model is not stagnant but rather continuing to evolve. After reading this chapter, you should have a basic understanding of these nontraditional cooperative forms and understand how they differ from traditional open-membership cooperatives.

CHAPTER 5

Basics of Cooperative Accounting

Introduction

In order to understand the financial condition and financial performance of a cooperative—or any business—one must understand the basic accounting concepts and be able to read financial statements. You also need to be able to identify which financial statement to review to answer a particular financial question. It is also useful to have a general idea as to how the process of operating the business affects its financial statements.

The financial statements for cooperatives follow generally accepted accounting principles (GAAPs). This means that in many respects, they are similar to the financial statements of other businesses. There are unique aspects of the cooperative model that create different terms and structures. For example, the statement of operations for a cooperative shows "profit before patronage" and then patronage distributions before arriving at "Net profit after patronage and taxes." Those patronage calculations would not be found on the financial statements of an investor-owned firm.

After reading this chapter, you should be able to understand the basic financial statements of a cooperative. A glossary is also available for quick definitions for unfamiliar terms and can be accessed at any time.

The Audit

Most cooperative bylaws specify that the organization will hire an outside auditor to conduct an annual audit. The audit helps to prevent deliberate misstatement of fact. It also ensures that judgment decisions are not duly biased in favor of management and that records are dependable. The audit also ensures that GAAPs have been consistently followed and the disclosure is complete.

The audit consists of a review of the following three basic financial statements:

1. Balance sheet
2. Income statement
3. Statement of cash flows, also called the cash flow statement

Many auditors also include the statement of members' equity.

The auditor examines the financial statements and their supporting documentation provided by the company and assesses whether the statements fairly present the financial condition of the cooperative. The supporting documentation might include:

1. A verification of accounts receivable and payable balances with cooperative customers
2. A review of inventory quality, quantity, valuation, records, and procedures

3. A verification of the existence of recorded securities
4. A review of the justification for judgment decisions and estimates
5. A review of a sampling of accounting records
6. A review of the minutes of the board of directors' meetings for policy changes and instructions to management.

After the audit is completed, the auditor issues a written report containing their opinion to the board of directors. This report takes a standard form, and various expressions confer the seriousness of any problems identified by the auditor. Auditors have the option of issuing four types of opinions. These opinions are as follows:

1. **Unqualified opinion:** Financial statements present fairly the financial position and operating results of the company in conformity with GAAPs. In effect, the auditor is stating that the audit is accurate and complete and that the auditor is placing no qualifications on their opinion that the financial statements are accurate. **The unqualified opinion is what members, chief executive officers, and boards of directors want to see.**
2. **Qualified opinion:** Statements are fair, "subject to" or "except for," because some area could not be tested or there was a situation that required deviation from the normal procedures. This means that the auditor is putting some qualification on their statement that the audit records are accurate. This does not necessarily mean that the accountant thinks something unethical is happening. It could, however, mean that the accountant found instances in which expenses should have been assigned to a different category, or there were some errors found in line items, such as transposed digits. The auditor is placing some qualification on their opinion that the statements are accurate.
3. **Disclaimer of opinion:** The auditor was unable to render an opinion on the fairness of the financial statements. This may mean that the records supplied were insufficient to prepare proper audited financial statements or that there were a number of issues that would have to be addressed before the accountant would be able to evaluate the accuracy of the information provided. Generally, when an accountant declines to issue an opinion, there is a need for the cooperative to retool their internal accounting procedures so that they can operate according to the usual and proper accounting standards.
4. **Adverse opinion:** The auditor feels the statements are not presented fairly, and the situation remains unresolved.

Audit Statements: Balance Sheet, Statement of Operations, and Statement of Cash Flows

Balance Sheet

PURPOSE AND ROLE OF THE BALANCE SHEET

The balance sheet is a snapshot taken at a moment in time. In a grain cooperative, the balance sheet would show the amount of grain inventory (an asset) on the day the audit was conducted. Usually, a balance sheet is produced at the end of a reporting period so that it corresponds to the other two types of reports produced at that time.

In the balance sheet, you can easily see the relationship between debt (i.e., liabilities) and equity. If these are equal—for every dollar of debt we have a dollar of equity—this would be a 1:1 debt-to-equity ratio. Where did all the funds for all the cooperative's assets come from? Half came from the lenders, and half came from the cooperative members. However, we can say that only for the one moment when the balance sheet was produced—when the snapshot was taken. Look for the single date on a balance sheet to see when this occurred.

It's handy to think of balance sheets as a sequence in a photo album that enables you to compare pictures over time. Think of a photo album of a child growing up with pictures taken once a year. If you compare the photos from year one, three, five, 10, and 20, it's easy to see that things have changed. It's the same with a balance sheet: Financial conditions change over time, and you can see the change by comparing balance sheets.

The balance sheet performs many functions, including:

- Outlining the financial position of a company at a specific point in time
- Summarizing what a company owns and owes to others
- Identifying what proportion of the company's assets are owned by the patrons and what proportion are (in essence) owned by creditors

The basic accounting equation is assets = liabilities + equities.

Everything an organization owns (i.e., assets) must be financed through owner equity or debt. For this reason, assets must equal liabilities plus owner equity. A balance sheet, in order to be accurate, must literally "balance."

In other words, we have a bunch of stuff, we owe money for some of it (liabilities), and the rest came from money the members paid plus money the cooperative earned (retained equity). Following is the simplified balance sheet.

Assets are the things the organization owns. They are usually organized into three categories: current, noncurrent, and fixed assets.

Current is generally defined as within the next 12 months. *Current assets* refers to cash, inventory, and any other assets that would ordinarily be or become liquid, or cashlike, in the next 12 months. Cash, accounts receivable, inventory, prepaid expenses, and advance payments are examples of current assets typically found on cooperative balance sheets. Accounts receivable and inventory are current assets because we would ordinarily expect them to be turned into cash within a year. That doesn't mean there are not rare cases when a cooperative holds inventory for more than a year or has to wait for that amount of time to collect an account receivable.

Noncurrent assets are items that cannot be converted into cash within one year's time but don't fit our definition of fixed assets. A common example of noncurrent assets is investments in other cooperatives. This asset occurs when the cooperative received patronage from a regional cooperative in the form of stock. The regional cooperative will eventually redeem the equity, so it is not a fixed or long-term asset, but the regional equity won't be redeemed within a year.

Fixed assets include buildings, facilities, land, vehicles. and equipment. They are shown on the balance sheet at their book value. The book value is calculated by taking the original purchase cost and subtracting the accumulated depreciation.

Statement of Operations

Also referred to as an ***income statement***, *profit and loss statement*, or *operating statement*, the statement of operations is a report that shows sales, the cost of sales, expenses, and profit for a period of time (e.g., a week, month, quarter, year) and then starts over for the next reporting period. In a grain cooperative, the statement of operations would show the amount of grain purchased and sold during a period of time. So, while the balance sheet is a snapshot at a point in time, the statement of operations is a summary during a period of time. This statement, which is generally included with the audit, covers the previous operating year. The statement of operations presents all the information that is necessary to get to the bottom line—in other words, the net profit. The bottom line on a cooperative income statement is often labeled *Net Savings* instead of *Net Income*. The statement may also show the distribution of the savings (net income) into cash patronage refunds, stock patronage refunds, and additions to unallocated reserves.

Some cooperatives are themselves members of regional cooperatives and receive patronage from these cooperatives. This shows up on the income statement as it creates the difference between local net savings and total net savings.

The basic equation of the statement of operations is as follows:

Sales – cost of sales = gross margin – expenses = local net savings + patronage received from regional cooperatives = total net savings

Statement of Operations

Jan. 1, 20XX, through Dec. 31, 20XX

Gross Sales	
Fertilizer	$4,800,000
Feed	$1,000,000
Wheat Marketing	$2,500,000
Wheat Storage	$96,000
Total Sales	$8,396,000
Cost of Goods Sold	$7,198,000
Total Gross Margin	$1,198,000
Operating Expenses	
Variable	$399,900
Fixed	$391,817
Other	$130,000

Total Operating Expenses	$921,717
Local Profit	$276.283
Regional Patronage	$100,000
Total Profit Before Patronage	$376,283
Cash Patronage Refund	$188,141
Before-Tax Savings	$112,885
Tax	$56,442
Net Savings After Tax	$56,442

Note that our example cooperative had $100,000 of patronage (income) from a regional cooperative and paid a portion of its savings to members in the form of cash patronage. The patronage distribution is based on total savings, so it passes on a portion of both the local and regional savings.

Statement of Cash Flows, Also Called the *Cash Flow Statement*

The cash flow statement is a summary of how money flows into and out of the organization. It is like a bank reconciliation statement—it summarizes all cash activities. There is a more expanded statement called *changes in financial condition*, but we will stick to the cash flow statement for our discussion.

While a cooperative must be profitable over the long run to survive, in the short run, it is essential that it has cash to pay its obligations. There are a number of reasons why profitability can be different from cash flow. A simple example would be redeeming equity (stock) that was issued in a previous year. This reduces the cooperative's cash flow. Equity redemption is not an expense, so it has no impact on profits. On the other hand, distributing profits in the form of stock can result in having a cash flow higher than profit. Certain types of stock patronage distributions are deductible to the cooperative, so those stock distributions reduce the before-tax profit but have no effect on cash. Loan principle payments and the purchase of additional assets are other common events that reduce cash flow but are not classified as expenses and do not affect profitability. Selling fixed assets is also a source of cash but is not income.

Depreciation is an expense that attempts to measure the reduction in the value of fixed assets that occurred during the period. Depreciation does not actually require cash, so it can cause cash flow to be higher than profitability.

In order to visualize the treatment of depreciation, imagine that you bought a lawnmower for your mowing business. If the mower cost $100 and lasted exactly four years with no salvage value, your statement of operations would show a $25 depreciation expense each year. After adding information on the fees you collected from mowing minus your other expenses you could determine whether you were making a profit mowing lawns. Your cash flow statement would show a cash outflow of $100 in the initial year and no cash effects during the remaining year. The depreciation expense on the statement of operations is a better reflection of the fact that you used up some of the value of the lawnmower each year. The cash flow statement shows when you actually needed the cash. Cash flow is $75 lower than profitability in the initial year and $25 higher for the next three years.

Statement of Patron Equities

The statement of patron equities is a supplemental statement that, although not required, is often included with the audit financial statements. The statement summarizes equity that all patrons jointly own in the company and outlines the portion of equity that is in the cooperative's name. It outlines additions to equity based on current profits, reductions to equity due to stock retirements, adjustments to unallocated equity, and other additions or subtractions to the equity accounts.

The basic equation for patron equities is as follows:

Beginning allocated equity
+ equity sales to new members
+ additional equity patronage issued
– allocated equities retired
= ending allocated equities

Supplemental Statements

Most cooperatives distribute the audited financial statements at their annual meeting or otherwise make them available to members. Many cooperatives prepare a variety of supplemental statements, for the long form of the annual financial statements and for internal use. Common supplemental statements would include the changes in financial condition, the statement of operations broken down by department, more detailed expense statements, and a schedule showing the aging of the accounts receivable.

Statement of Changes in Financial Position

The statement of changes in financial position, also called *funds flow* or *changes in working capital*, shows where operating funds are coming from and where these funds are being used. It is more general than the cash flow statement because it indicates changes in working capital and not just those in cash. As a simple example, if a cooperative uses cash to buy inventory, its cash balance is reduced, but its current assets and working capital have not changed because inventory is still a current asset. Most lenders would not be concerned that the current assets now consisted of inventory instead of cash because they would expect the cooperative to soon sell the inventory and convert it to cash. You can see from this example why the statement of changes in financial condition, which tracks changes in working capital, might be a more useful statement.

The basic equation for the statement of changes in financial position is as follows:

Sources of funds – uses of funds = change in working capital

Working capital is calculated as current assets – current liabilities = working capital.

Working capital measures whether the cooperative has enough assets that it can convert to cash to pay the bills and obligations that will require cash. Organizations must maintain enough working capital to pay bills, purchase inventory, and finance accounts receivable (credit to patrons). While managers must forecast cash flow in order to be sure they will have sufficient cash, lenders tend to be more concerned about the cooperative's working capital.

Summary

The financial statement of a cooperative, as generated by the audit process, contain the following three very important categories of information:

1. The balance sheet shows what the cooperative owns (assets), what it owes (liabilities), and how much the members own (equity) at a point in time.
2. The statement of operations shows the revenue, expenses, and profit and loss for the period of time covered by the statement. It essentially shows the financial performance of the cooperative.
3. The cash flow statement shows how cash flows into and out of the cooperative and the cash situation. Cash flow is very different from profit and loss, so it is another important aspect of the financial performance and condition. The net effect of how owners' equity changed can be seen by comparing the subsequent balance sheet. The details of how equity is created and revolved are shown in the statement of patron equity.

Glossary of Financial Terms

Accounts receivable (receivables)
Money due to the cooperative from the sale of goods and services on credit. Amounts shown are net of allowance for doubtful accounts (if reserve method of accounting is used). Accounts receivable are a current asset.
Accounts payable (payables)
Money payable for the purchase of goods on credit; the amount the company owes its suppliers and vendors. Accounts payable are a current liability.
Accruals
Amounts set aside to pay for bills known but not yet received. Examples include insurance, taxes, and payroll. Accrued expenses are a current liability.
Adverse opinion
The auditor feels the statements are not presented fairly, and the situation remains unresolved.
Assets
The entire property of all sorts belonging to or in the possession of a corporation; includes current assets, fixed assets, and investments.
Balance sheet
Shows what the cooperative owns (assets) and how those assets are financed (debt and equity) at a point in time.
Cash
Currency and funds on deposit in checking or savings accounts.
Cash patronage
The portion of total patronage paid in cash.
Common stock
Patron ownership in the organization; earned through doing business with the cooperative. Common stock does have voting rights.
Current assets
Assets that can be converted to cash through the regular course of business within the next 12 months; typically include cash, accounts receivable, and inventory.

Current Liabilities
Amounts that must be paid to outside parties within the next 12 months; typically include accounts and notes payable.
Current ratio
Indicates the extent to which short-term debt obligations are covered by current assets; calculated by dividing current assets by current liabilities.
Depreciation
Most major assets are not used up or worn out within a one-year period. *Depreciation* is the process where a portion of the value of an asset is set aside each year so that at the end of the asset's useful life, the cooperative has funds to replace the asset. This is a commonly accepted business practice. The operating statement or expense report will show the amount of depreciation expense that occurred during the period. The balance sheet may show the value of the fixed asset net of all the accumulated depreciation or show the asset at their original value followed by an offsetting entry for accumulated depreciation.
Disclaimer of opinion
The auditor was unable to render an opinion on the fairness of the financial statements. This may mean that the records supplied were insufficient to prepare proper audited financial statements or that there were a number of issues that would have to be addressed before the accountant would be able to evaluate the accuracy of the information provided.
Equity credits
See *Stock credits.*
Equity retirement (also called *equity revolvement* or *stock revolvement*)
The process in which older stock is removed from the cooperative's books and its value is paid out to patrons. Stock revolvement allows cooperatives to convert patrons' equity into cash (after a period of time established by the cooperative). It also keeps the cooperative's ownership in the hands of more recent patrons.
Expansions
Significant additions of property, buildings, equipment, and/or other business assets for the purpose of growing sales and profits.
Financial accounting standards board
Seven-member accounting industry board that issues authoritative expressions of GAAPs.
Financial ratios
Quantify the relationships between various financial elements; allow consistent and reliable analysis of financial statements as well as comparisons between cooperatives.
Fixed assets
Investments in land, buildings, equipment, fixtures, and other tangible assets required to conduct the business.
Gross margin
The dollars available from the sale of goods and services after subtracting the cost of purchasing the goods or services from suppliers.
Independent auditor
Accountant, typically a certified public accountant, who is not directly or indirectly affiliated with the cooperative and who conducts a thorough and complete review of the cooperative's records and accounting practices.
Inventory
Goods on hand available for sale through the cooperative.
Liabilities

Amounts the cooperative owes to suppliers, banks, employees, and the government (taxes).

Local net savings

Earning generated by the local cooperative for a specific period of time; does not include patronage received from other cooperatives. Also called *local net profit*.

Long-term debt

Portion of bank loans due and payable for periods longer than one year; typically used to finance fixed assets or cooperative growth.

Long-term debt ratio

Quantifies the relationship between long-term debt and owners' equity; indicates what portion of the cooperative is owned by debtors and what portion is owned by members.

Marketing cooperatives

Cooperative businesses organized and operated to assist producers in profitably selling their commodity products. Many marketing cooperatives such as Land O'Lakes and CHS cooperatives add value to producers' products by processing them into consumer products.

Member patrons (patrons)

Customers who have earned equity in a cooperative by purchasing products or selling commodities through the cooperative.

Noncurrent assets

Assets that, through the regular course of business, are not expected to be turned to cash within a 12-month period; include investments in other cooperatives. Fixed assets such as land, building, and equipment may be classified as noncurrent assets or in a separate category of fixed assets.

Noncurrent liabilities

Amounts owed and payable at a time period more than 12 months into the future.

Owners' equity

See *Patron equities*.

Ownership ratio

Indicates what proportion of the cooperative is actually owned by member patrons.

Patron equities

The amount of the cooperative that patrons own; includes all stock and reserves; considered the organization's net worth.

Patronage

Distribution of earnings from a cooperative to its members/owners; includes both stock patronage (increased ownership in the cooperative) and cash amounts.

Patronage refunds payable in equities

The process of the cooperative distributing a portion of the patronage refund in the form of stock. The cooperative issues additional stock to the individual members in proportion to their patronage. This additional equity is reflected on the balance sheet and statement of patrons' equity. Stock patronage refunds can be paid in the form of qualified or nonqualified stock.

Prepaid expenses

Items that are assets at the time of purchase but become expenses as they are consumed. Examples include insurance premiums, office and store supplies, prepaid rent, prepaid taxes, and prepaid wages.

Qualified opinion

Statements are fair, "subject to" or "except for," because some area could not be tested or there was a situation that required deviation from the normal procedures. This means that the auditor is putting some qualification on their statement that the audit records are accurate.

Stock credits

Many cooperatives do not issue physical shares of stock or put a specific value (e.g., $1) on a share of stock. The member's stock balance is maintained as stock credits or book credits. This eliminates the problem of lost or misplaced certificates. With stock credits, a member can be issued any amount (e.g., $15.50), which would not be an even number of shares if the cooperative uses certificates and a stated share value.

Stock retirement (stock revolvement)

The process of the cooperative issuing cash in exchange for the stock or stock credits that were previously issued as a stock patronage refund. Stock revolvement is necessary because there is generally no market for the cooperative stock and the member realizes a benefit from the stock patronage refund only when it is eventually redeemed for cash. Most cooperatives have established systems for stock revolvement which may be based on the age of the stock, the age of the member, or special situations. Some cooperatives redeem a percentage of all members' equity or use a base capital program that matches investment with use.

Supply cooperative

Cooperative businesses organized and operated to assist producers in acquiring the products and services they need to run their farm of business profitably.

Unallocated retained earnings (also called *unallocated equity* and *unallocated capital reserve*)

Equity in the cooperative that is not attributed to specific members. Most cooperatives channel a portion of their profits to unallocated equity in order to create a reserve to guard against future losses. Earnings from nonmember business or gains on sales of assets may also be channeled to unallocated equity. This account provides a cushion fund that can be reduced when the cooperative experiences losses. In the absence of unallocated equity, a cooperative must write down the value of member stock when it experiences a loss.

Unqualified audit report (unqualified report)

On completion of an audit, the auditor must express an opinion as to whether the financial statements are fairly presented or issue a disclaimer of opinion which states that they cannot issue an opinion. In an unqualified audit report, the auditor states that they are completely satisfied that all required procedures have been performed and the statements as presented are without major errors.

Working capital

Monies needed to fund the day-to-day operation of the business; used to pay bills, purchase inventory, and finance accounts receivable; calculated current assets minus current liabilities.

CHAPTER 6

Economic Justification for a Cooperative

Introduction and Learning Objectives

Individuals often form or join cooperatives to accomplish a goal collectively that they could not achieve individually. Some cooperative businesses are particularly successful because their structure gives them an advantage over other business forms. Both of those outcomes reflect the fact that there can be economic justifications or economic rationales for a cooperative. Historically, most U.S. agricultural cooperative were formed to offset monopoly power and create economies of scale. Those are just two examples of economic justifications of cooperatives. In this chapter, you will learn more about those justifications and the other economic forces that can make a cooperative the logical solution to fulfill a need.

Why Are Cooperatives Formed?

In one sense, the cooperative form of business needs no more justification than do proprietorships, partnerships, or investor-owned corporations. Businesses are formed when the owners perceive a positive return on investment from providing some combination of products and services to customers. A cooperative is simply an alternative structure for organizing economic activity. The major rationale for forming a cooperative is to improve the economic well-being of the member-owners. In almost any situation where there is a potential to form a feasible business, there is the potential to do so as a cooperative, and in fact, cooperatives successfully operate in almost every business sector. However, there are situations in which the cooperative form of business is particularly effective. This is especially true in the case of agricultural cooperatives.

Economies of Scale

When the per-unit cost of an operation decline as the number of units increases, there are positive economies of scale. Economies of scale are a classic rationale for starting and operating a cooperative. The per-bushel cost of a 1 million–bushel grain storage structure is typically half that of a 10,000-bushel facility. It is easy to see why producers might want to join together to form a grain storage or fertilizer warehousing cooperative since the cost of jointly owning and operating those functions would be lower than investing in on-farm grain and fertilizer storage. Scale economies also often extend to purchasing and marketing. Sourcing inputs or marketing commodities requires someone to constantly monitor the markets and buy or sell at the appropriate time. The per-unit cost of monitoring and making those transactions declines as the volume being handled increases. The creation of economies of scale is one of the rationales for almost every cooperative, although it is more obvious in some situations than in others.

Missing Services

Entrepreneurs are always searching to identify a product or service that customers are willing to purchase at a price that will cover the cost of production and provide a reasonable return on the invested capital. The perception of a missing service can be a rationale for establishing a new business. Many cooperatives are formed to provide a missing service. In many cases, that service is somewhat specialized or focused in terms of a particular group of users and has not caught the attention of the broader investment community. Rural electric cooperatives (RECs) began forming in the late 1930s to bring electricity to rural America. Investor-owned utilities were not interested in investing in electrical infrastructure in rural areas because the customer density was much lower relative to urban areas. Today's REC system, which controls 44% of the lines in the United States, was clearly created to provide a missing service. Another example of offering an absent service was the establishment of organic food stores in the early days of the organic movement. The customer base for organic foods was too small to interest the general business community. That led organic food consumers to form cooperatives to provide the missing service of retail organic outlets. As the demand for organic foods has grown, those cooperatives continue to prosper but face competition from traditional retail outlets that have now expanded into organic product lines.

Risk Reduction

Every consumer and business faces risk, and it is particularly prevalent in agriculture. Some risks can be eliminated, but in many cases, the only solution is to share the risk with a pool of other individuals or businesses. Risk pooling is the sole purpose of insurance cooperatives. Like all insurance companies, cooperative insurance companies allow policyholders to pool risk and pay a share (i.e., the premium) of the risk each year while avoiding a catastrophic risk. Because the policyholders in an insurance cooperative are owners and receive the residual profits, they have a built-in incentive in controlling risk. For that reason, many cooperative insurance companies have active and effective programs to help their owner-members reduce and control loses.

Many other types of cooperatives have risk pooling as part of their functions, and many agricultural marketing cooperatives have marketing pools. These allow the pool participants to receive the average annual price for the grade and quality of commodity they placed in the pool. Marketing pools eliminate the risk of selling at a below-average price at the cost of missing the opportunity to sell at an above-average price. Many cooperative members conclude that they cannot beat or time the market and choose to eliminate their market price risk through the marketing pool. Other cooperatives reduce risk through their everyday operation. If a farm supply and marketing cooperative chooses to operate with a single patronage (profit distribution) pool, the risks of losses in agronomy or grain handling are shared. The basic structure of shared services also pools risks. If a producer operates their own sprayer or fertilizer applicator, they are responsible for the entire risk of any breakdowns. When those services are provided by the cooperative, the cost of repairs is implicitly shared across all users through the application fee. The cooperative member using the service does not bear the cost of the applicator breaking down on their job.

Unique Role of Agricultural Cooperatives

As mentioned, there are some rationales for a cooperative that are somewhat unique to or at least particularly prevalent in the agricultural industry. Agricultural cooperatives often function as an extension of the farm firm.

The cooperative extends the farmers' business backward, into the input supply chain, or forward, into the marketing and processing value chain. This extension of the farm business is referred to as *backward integration into the input supply chain* or *forward integration into the marketing chain.* As we will see later, this concept of the agricultural cooperative serving as an extension of the farm business is the basis of cooperative taxation and the rationale for allowing cooperatives to pass the taxation on distributed profits on to the farmer members. This unique role of the agricultural cooperative also creates challenges for the manager and board. Patrons are affected by both the price paid or received and from the share of the cooperatives profits. Increasing profitability by purchasing a member's commodities at a lower price does not benefit the member because the increased profits came from reduced commodity payments. The cooperative board and manager must focus on increasing the profitability of the combined farm and cooperative system.

Markets Are Not Perfect

If producers faced the perfect markets that you studied in your introductory economic class, they might not be as interested in starting cooperatives. For example, if markets were perfect, every cotton farmer would receive the same price for their cotton, adjusted only for transportation. In the early days of cotton farming, cotton was transported loosely in horse-drawn wagons. That gave the producer no practical choice as to market outlet since it was impractical to haul to another town. That gave each cotton gin a mini-monopoly. Not surprisingly, a strong cooperative ginning industry developed. Incidents of market power were quite common in the early days of U.S. agricultural due to the dispersed nature of commodity production.

Offsetting Monopoly Market Power

A primary characteristic of a competitive market is the existence of a large number of buyers and sellers. In that situation, no market participant has the ability to influence the market price. Fertilizer manufacturing, petroleum refining, food processing, and many other agribusinesses are all characterized by a relatively small number of firms on one side of the market. Much of this is due to economies of scale. A large flour mill or large fertilizer manufacturing plant has a lower cost per unit relative to a small one. When scale economies exists, larger firms grow, and smaller firms go out of business or merge. That leads to an industry structure with a limited number of firms. When there are a limited number of firms in an industry, those firms can have market power (i.e., they have the ability to influence the market price). The most extreme case is that of a monopoly (one supplier) or a monopsony (one buyer). The need to offset market power in markets with a small number of buyers or sellers is a classic rationale for the formation of a cooperative.

Most of our legacy agricultural cooperatives in the United States were established in the 1920s and 1930 during a period of time when large banks and railroads had high levels of market power in rural America. The Capper-Volstead Act, which provides the legal foundation for U.S. agricultural cooperatives, was a direct result of anticompetitive environment during that period. For that reason, the importance of offsetting market power as a rationale for cooperative formation cannot be overemphasized.

Distance and Transportation Also Enter In

In agricultural markets, there are also often spatial dimensions to the market, or *spatial economic forces*. When the cost of transportation is high relative to the value of the commodity or the product is perishable or difficult to transport, there can be limited competition in a geographic area. For example, a dairy farmer might have only one logical to market their milk even when there are multiple milk processing plants in the United States. This is a common situation in many rural areas, particularly in the Great Plains states. In this case, the buyer or seller is characterized as a spatial monopolist and has market power within a geographic area.

There is often a second dilemma in markets with limited competition within a geographical area. Adding more buyers or sellers might address the market power issue, but the least cost structure in terms of manufacturing and transport costs might involve a single buyer or seller. For example, if multiple milk processing plants operate within a county, they likely have overlapping routes picking up the producers' milk. The plants will also be smaller than the most efficient scale and therefore have a higher cost. The most economic structure for transport and processing might involve a single firm. However, the existence of a single milk processor, organized as an investor-owned firm, raises the danger of market power. The milk processor would have a spatial monopoly and could use their market power to offer the producers an unfairly low price. A cooperative milk processing plant provides the ideal solution. It allows the producers to gain the transportation and scale economies of a single large processing plant without concerns about market power. Because the cooperative returns all its profits to the members supplying the milk, it has no incentive to offer unfair prices. Cooperatives have a very large market share in dairy processing, which reflects the fact that producers understand the economic rationale that a cooperative can provide the scale economies of a single plant without concern over unfair pricing.

Transaction Costs

You might recall that in our perfect market, every participant has perfect information, and transactions are costless. The reality of many agricultural markets is that information is imperfect, and transaction costs can be significant. A producer may not have access to the prices at all possible outlets, and getting current bids may involve both a monetary and time cost. The lack of complete price information can be a particular problem for fruit and vegetable producers and other farmers with perishable products. The producer has a limited time to market their commodities and may sell at an unfavorable price due to limited information. Providing price information and facilitating transactions is a key rationale for many specialty crop marketing cooperatives. There is a successful apple marketing cooperative in New York that supplies only market price information. Their members make the actual transaction and package and transport the product. The cooperative still increases member profits by more than 10% simply by giving the producers complete access to prices at all possible outlets.

Bulk commodity marketing cooperatives still play a role in reducing transaction costs. Individual grain producers would find it expensive and time-consuming to meet the contract specifications of a flour mill or to export their grain. The grain marketing cooperative reduces those transaction costs by marketing for the combined membership.

Information can also be a part of the value of a commodity, and cooperatives can create value by helping certify and communicate that information. For example, the grade for a product may not represent all the quality characteristics. A carton of eggs might be graded "Grade A Large"; however, the eggs might also come from chickens raised in free-range conditions. A steak graded "Choice" might originate from grass-fed beef.

Cooperatives such as Organic Valley have created value by conveying information to consumers. The Organic Valley brand conveys information about how the milk was produced, which is important to some consumers.

Coordination

Most agricultural products are processed or at least handled and packaged before they reach the final consumer. Processing operations have high fixed costs and are most efficient when operated at a constant volume that is at or near their maximum capacity. This creates the challenge of coordinating production with processing. As an example, consider a cull cow processing plant that supplies ground beef to hamburger chains. Ideally, the plant would operate at full capacity year-round. Unfortunately, most beef producers cull cattle seasonally. That creates a challenge for the plant to procure a constant year-round supply.

Theoretically, a processor can solve the problems of coordination through contracts and date-based premiums and discounts. Contract coordination is common in some aspects of agricultural such as broilers and hogs. Contracts can also be risky and expensive. Also, since there is often a single processor dealing with many producers, the producers may fear an imbalance of market power. Having the producers forward integrate into the processing operation has the potential to achieve gains from coordination. Because they share in the resulting profits, the producers are often much more willing to match the timing of delivery with the processing plant's needs. There can also be further potential gains from producing the crop or livestock varieties and sizes that lowers processing costs and/or creates the highest value in the final product.

Access to the Market

Another rationale for the formation of agricultural cooperatives is simply to have access to the market. The least cost shipping bulk commodities to many markets involves transportation by train or barge. Individual producers do not have rail or river access or the volume necessary for a full train or barge shipment. Food processors such as flour mills have strict quality standards and require sellers to have large liability insurance policies. A typical flour mill purchases more than 1 million bushels a week. Flour mill buyers are therefore not interested in dealing with suppliers that cannot provide that level of volume on a weekly basis. Similarly, a grocery chain may purchase millions of eggs per year. They will purchase only from suppliers that can meet extensive food safety standards and supply their needed volumes. In all these situations, marketing cooperatives provide producers with the ability to access the market. The cooperative has the scale of operation, food safety expertise, infrastructure, and expertise to access markets that are inaccessible by the individual producer.

Concepts of Competitive Yardstick and Invisible Benefit

A famous cooperative scholar, Edwin Griswold Nourse, developed what is known as the ***competitive yardstick*** school of thought on the role of cooperatives (Schomisch, 1979). According to this concept, a cooperative's role is to provide a measure of reasonable prices by operating at cost (including a normal return for capital invested). The competitive yardstick model implies that a major role for cooperatives is to maintain competitive and efficient markets. This leads to a related concept that one advantage of a cooperative is its **invisible benefit** in keeping the market competitive. Unfortunately, that benefit is often not recognized until the cooperative disappears and the remaining firms are free to exercise market power through unfavorable prices.

An interesting example of the competitive yardstick comes from RECs. In most states, electric utilities are regulated by the state's corporation commission or similar entity. The assumption is that in the absence of regulation, the utility would use its monopoly power to charge unfairly high rates and generate excessive profits. RECs are generally exempt from regulation. The reason for the exemption is because of their cooperative business model, if an REC were to charge excessive rates, it would simply refund the excess to the same customers through patronage refunds. For that reason, regulators consider RECs to be self-regulating. One could also say that if we wanted a measure of the fair and reasonable price for electricity, we could look to RECs for a competitive yardstick.

Summary

Cooperatives are an alternative form of organizing a business. Their basic rationale is that they provide a good or service that is demanded at a price that creates an acceptable return to the member-owners. There are, however, some classic situations in which cooperatives are particularly effective. In terms of the general cooperative business model, typical rationales include creating economies of scale, providing a missing service, and reducing or pooling risk. Those rationales also apply to agricultural cooperatives, which operate as an extension of the farm firm and also often in imperfect markets. Some of the classic rationales for the formation of agricultural cooperatives include offsetting market power, addressing the spatial dimension of many agricultural markets, providing market information, reducing transaction costs, and coordinating production and processing. Most U.S. agricultural cooperatives were formed to offset market power, and cooperatives continue to have a role in providing a competitive yardstick as to fair and reasonable prices and generating an invisible benefit in keeping markets competitive.

Reference

Schomisch, T. P. (1979). *Edwin G. Nourse and the competitive yardstick school of thought* (University of Wisconsin Center for Cooperatives, Occasional Paper No. 2). University of Wisconsin.

CHAPTER 7

Cooperative Legal Environment

Introduction and Learning Objectives

In terms of their legal identity, U.S. cooperatives are corporations. Cooperatives are incorporated under separate statutes, which gives them distinct legal features. This makes it important to understand the legal environment under which cooperatives operate. In addition, it is important to understand the hierarchy of legal documents that control all corporations. There are also some important areas in which the legal characteristics of cooperatives differ from those of other corporations. This chapter will help you understand all the elements of a cooperative's legal environment.

The Legal Structure of Cooperatives

As mentioned, cooperatives are a special form of corporation. A *corporation* is a separate legal identity that has rights and responsibilities just like a natural person. A corporation can be chartered with unlimited life, loan and borrow money, enter into contracts, and sue and be sued. An important characteristic of a corporation is that of limited liability. Shareholders have the right to share in profits and losses but are not held personally liable for the corporation's debts. Corporations are regulated under the corporate laws of the state in which they are incorporated. Corporate law deals with the formation and operations of corporations and the regulations they must follow to enjoy the tax and legal benefits of corporations.

The majority of the corporations in the United States are referred to as *C corporations* (C corps). In a C corp, the business' profits are taxed at the corporate level, and any dividends distributed to the owners are taxed again as part of the shareowners' taxable income. A number of these corporations are publicly traded, which means that they issue stock that can be bought and sold in an organized stock exchange. The owners of publicly traded companies include institutional investors, such as mutual funds and pension funds. When a publicly traded corporation retains funds and invests it in profitable projects, this action tends to make the stock price increase. The shareholders of C corps share in the firms' profits in proportion to their ownership, either directly through dividends paid in proportional to ownership or indirectly through stock price appreciation. As mentioned, cooperatives are a special class of corporation that use a different structure of profit distribution.

The Hierarchy of Documents

Three documents relating to the legal structure of a corporation are the articles of incorporation, bylaws, and policies. The articles of incorporation are the legal basis for chartering the corporation within the state in which it is incorporated. These articles contain general information about a corporation, such as the name, location, purpose, length of existence (which can be perpetual), and names of incorporating directors. The criteria for disposing of assets if the cooperative is dissolved are also usually specified in the articles. The articles of incorporation are signed by the incorporating directors and filed with the state. They describe the procedure for amendments, any of which require not only a vote of the membership but also a refile with the state. For that reason, corporation tend not to specify any structures in the articles of incorporation that could be addressed in the bylaws.

The bylaws contain information about the rules and regulations that govern the corporation, such as membership, voting procedures, annual meeting procedures, structure of officers, and board of directors. The bylaws often describe specific powers for shareholder action, such as the procedures for a group of members to call a special meeting or for removing a board member. The initial set of bylaws is voted on and approved by the membership at the first annual meeting. Amendments to the bylaws must then be approved by the membership under the procedures described in the current bylaws. While amending the bylaws is less burdensome than amending the articles of incorporation, it still requires specific procedures for notifying the members and a membership vote during the next annual meeting (unless a special meeting is called). Although rules and procedures of governance issues are often covered in bylaws, the rules of operational manners and some of the specifics for governance procedures are typically covered in corporations' policies.

A corporation's policies are formulated and passed by its board of directors. The membership does not generally vote to approve policies; however, the board may inform the members of key policies that affect them (e.g., the credit policy). Policies typically specify guidelines and procedures for dealing with foreseeable situations and reoccurring events. They may place limitations on certain actions or establish risk management procedures. Policies can help a corporation make a consistent response to a recurring situation. Policies can also help ensure that the corporation is complying with laws and regulations and generating the appropriate documentation. Although corporate laws do not require that a corporation have policies, they can be an important issue in legal disputes. In many disputes centering on whether the board of directors acted properly or treated a member or employee fairly, the first line of investigation will be whether they complied with the organization's policies.

Cooperative statutes tend to be brief and focus on the unique structural requirements of a cooperative corporation. When there are legal aspects not covered by the cooperative statutes, the provisions of the state's general corporate statues apply. As an example, in cases where a cooperative board member is alleged to have violated their duty of care to the company, the courts will evaluate the case based on the business judgment rule. Under this standard, a board member is not liable for any outcomes from their decisions on the board as long as the decision was made (1) in good faith without a conflict of interest, (2) with the care that a reasonably prudent person would use when faced with a similar decision, and (3) with the reasonable belief that they are acting in the corporation's best interest. This is just one example of how, when an issue is not specifically covered in the cooperative statutes, the courts look to general corporation statues.

In addition to statues specifically related to cooperative corporations, cooperatives face all the legal regulations that individuals and other forms of business face. These include environmental law, employer/employee law, credit and finance law, real estate law, immigration law, bankruptcy law, and many other areas of laws and regulations, many of which are extremely significant to cooperatives. Part of the oversight role of the board of directors is to ensure that the cooperative is in compliance with its own articles of incorporation, bylaws, and policies and with all other laws and regulations.

Legal Issues Handled Differently by Cooperatives Relative to Other Corporations

There are a number of legal areas that are handled differently by cooperative corporations. One of these areas involves antitrust regulations, which is the topic of the following chapter. Areas of differences include voting systems, member eligibility, board member eligibility, restriction of dividends on invested capital, distribution of losses, and the distribution of the residual value at liquidation. From a legal standpoint, cooperative members often do not have a greater right to information relative to stockholders in a general corporation. However, information is often more continuous in a cooperative.

Voting Systems

In most investor-owned corporations—and investor-owned firms (IOFs) in general—voting rights are allocated in proportion to ownership. A shareholder who owns the majority of the corporation's equity gains a controlling interest in the firm. In contrast, one of the fundamental principles that distinguishes a cooperative is the principle of democratic control. Voting rights in a cooperative are either allocated on a one-member-one-vote basis or, less commonly, in proportion to business volume. Voting rights are not allocated on the basis of ownership. It should be noted that the Internal Revenue Service (IRS) tends to favor the one-member-one-vote system when analyzing whether a business is operating on a cooperative basis and qualifies to be taxed as such. While the federal statutes providing limited antitrust exemptions for cooperative specify either one member, one vote, or a restriction on dividends, many state cooperative statutes strictly specify one member, one vote.

Issues Requiring a Vote of the Membership

As discussed, the articles of incorporation and bylaws generally specify the actions that require a vote of the membership. In general, members vote to elect the board of directors, on amendments to the articles of incorporation and bylaws, and on very significant structural changes, such as mergers or liquidation. While cooperative corporations do not generally have stricter requirements for membership (i.e., shareholder) votes relative to other corporation, the nature of the cooperative firm leads to more frequent questions about what issues should be taken to a vote of the membership.

For example, a cotton ginning cooperative in Texas decided to exit the agronomy business and sell off all its fertilizer warehouses, applicators, and farm supply stores. The board was concerned that the time and publicity of taking the issue to a membership vote would cause them to lose an attractive offer for the assets. For that reason, they made the decision at the board level. While the board of directors likely had the legal authority to make the decision, many members felt that the board had a moral obligation to take the issue to a membership vote. In a cooperative, in which the owners are also customers, there are more often issues as to whether member input should be obtained. On the other hand, in order to operate as an efficient business organization, the board cannot bring all gray-area issues to a vote of the membership.

Voter Eligibility

In an investor-owned corporation, any individual or corporation owning a share of stock is entitled to vote in proportion to their stock ownership. In a cooperative, voting eligibility can be more complicated. In the scenario where the cooperative uses a one-member-one-vote system, there can be questions as to whether a husband and wife get one vote or two. Similarly, if a farm operation also involves the children, questions as to their eligibility to vote and at what age arise. Most farm operations now operate as limited liability companies (LLCs) or family corporations. In the case in which there are four farm families operating under an LLC, defining who has voting rights is obviously an important issue. In some cases, voter eligibility issues are defined in the bylaws. In many cases, the bylaws specify that an individual must be engaged in agricultural production to hold a membership and have voting privileges. That creates another criterion for a cooperative to consider as it reviews the farm's business structure to define the voting members. Some cooperatives require a base level of patronage during a specified period to maintain voting privileges. The goal of these provisions is to ensure that control is vested with active users.

Director Eligibility

Most corporate statues allow any natural person to be a corporate director. The shareholders can elect virtually anyone to serve on the board. In practice, the existing board and chief executive officer have the greatest influence on who is nominated. That allows the corporate board to have a formal succession plan and identify people with specific skills (e.g., finance or legal expertise) to bring a specifically needed skill set to the board. In a cooperative, only a member of the cooperative can be elected to the board. In an agricultural cooperative, a board member must therefore be an agricultural producer who is a member of the cooperative. While the cooperative model brings customers into the boardroom, it also inherently limits the diversification and potential skill sets of directors. That is one of the reasons why board education is an important ingredient for cooperative success. Other ways to compensate for the lack of diversity in backgrounds include outside advisory members, who typically provide insights but have no voting rights. The use of consultants and reports from staff members can also help fill gaps in skills and analysis.

In regard to diversity, agricultural cooperatives have traditionally had a low level of gender diversity. Progressive cooperatives are making more efforts to identify, recruit, and nominate female board candidates. In order to achieve that goal, the cooperative must also consider its membership eligibility guidelines and ensure that potential candidates are listed on the membership/voting rolls.

Allocation of Profits and Loss

In an investor-owned corporation, profits and losses are distributed in proportion to ownership. An investor-owned corporation generally has the freedom to retain all its earnings to maintain cash flow and reinvest in equipment and infrastructure. In contrast, in a cooperative earnings and losses are distributed in proportion to business volume. If a cooperative elects to retain funds by issuing qualified stock (which gives it an immediate tax deduction), it faces the restriction that it must distribute at least 20% of profits as cash patronage. The IRS has also taken the position that a cooperative cannot retain excessive amounts of earnings as unallocated reserves. In recent years, the level of unallocated reserves (unallocated retained earnings) relative to the amount of allocated equity (stock held by specific members) has increased. Many cooperatives now have unallocated reserves that represent more than 50% of total equity. To date, the IRS has not challenged these equity structures. It is still safe to say that a cooperative has much less flexibility in retaining funds as unallocated reserves (retained earnings) than IOFs.

Neither in an IOF nor a cooperative do the owners directly reimburse the company for their share of the firm's losses. Losses are allocated to the owners by reducing the value of their ownership. Differences in the systems for allocation of losses between cooperatives and IOFs relate to which owners have their equity values reduced. When an IOF experiences a loss, the value of all the owners is effectively reduced in proportion to ownership in the firm. One function of unallocated reserves in a cooperative is to absorb losses. When a cooperative experiences a loss, the board may elect to absorb it by reducing unallocated equity. In that case, the treatment of losses is similar to that in an IOF. However, when the loss exceeds the amount of unallocated reserves or the board determines it is more equitable, the loss may be absorbed by reducing the value (i.e., writing down) the value of membership stock. In determining which stock to write down, the cooperative generally considers the amount of business volume done during the most recent fiscal year or during the period associated with the loss. This implies that in a cooperative, losses are allocated to the members who did business with a cooperative during some recent period and are allocated in proportion to use. In simple terms, in a cooperative, losses are distributed in proportion to past use, whereas in an IOF, losses are allocated in proportion to ownership.

Liquidation

A corporation (investor-owned or cooperative) can be dissolved using a process called *liquidation*. In liquidation, a liquidator is appointed, and the corporation's assets are sold. The value received for the assets is used first to pay the creditors in the order of their established claim. Any remaining value is distributed to the owners. The difference in liquidation treatment between IOFs and cooperatives relates to the basis by which the owners receive their share of the residual value. In an IOF, the residual value of a firm is distributed in proportion to ownership. In a cooperative, the residual value is distributed in proportion to business volume during a look-back period. This system of distributed the residual value favors active users. A common length for the look-back period is five to six years.

Dividends

In an IOF, there is no restriction on the amount of profits that can be distributed as dividends. (A *dividend* refers to a payment made in proportion to stock ownership. While the IRS uses the term *patronage dividend*, the preferred term for a profit distribution made in proportion to use is *patronage refund*.) In some corporations, particularly firms with stock that is not publicly traded, the receipt of dividends is the major avenue by which owners benefit from their ownership in the company. Some corporations with publicly traded stock also pay dividends. In other cases, the corporation retains the profits, which tends to increase the stock price. While not every IOF pays dividends on its stock, there are no restrictions as to the level of the dividends.

In a cooperative, dividends are often limited not to exceed 8%. This is referred to as *subordination of capital* and is intended to limit the returns on ownership so that the primary focus can be on distributing profits in proportion to use. The limitation of dividends to 8% was one of the original Rochdale principles and is still reflected in statutes affecting cooperatives. The specific limit of 8% is also referred to as the *bright-line limit on dividends.* The antitrust exemption legislation that enables cooperatives at the federal level includes a provision that a cooperative must either limit dividends to 8% or use a one-member-one-vote system. Many state statures include both requirements and thus have a firm limitation of 8%.

The IRS tax code for Section 521 cooperatives—a more restrictive structure of cooperatives that qualifies for more favorable tax treatment and exemption from security registration—also has the 8% dividend restriction. In regards to taxation, it should be noted that in a cooperative, patronage refunds (profit distributions in proportion to business volume) are deductive from the cooperative's taxable income. Dividends (profit distributions based on ownership) are generally not deductible. A cooperative distributing profits as dividends must therefore make the payment out of its after-tax profits, and the dividends received by the cooperative member are essentially taxed twice, once at the cooperative level and again when they are added to the member's taxable income. While this tax treatment is a separate issue from the restriction on dividends, it does explain why the practice of paying dividends on member stock is uncommon in cooperatives.

Summary

Cooperatives are a special form of corporation. As such, they must act within the structure of their articles of incorporation, bylaws, and policies to maintain their treatment as a separate legal entity. A board operating outside of the cooperative's structure, or one that does not exercise an acceptable duty of care, risks being held personally liable for errant decisions. Cooperative statutes tend to be brief and focus on the unique structural requirements of a cooperative corporation. When there are legal aspects not covered by the cooperative statutes, the provisions of the state's general corporate statues apply. Cooperatives, like other companies, must also comply with a wide range of general laws and regulations. There are some key legal areas that are treated differently by cooperatives relative to investor-owned corporations. These include voting systems, member eligibility, board member eligibility, restriction of dividends on invested capital, distribution of losses, and the distribution of the residual value at liquidation.

CHAPTER 8

Cooperatives and Antitrust Legislation

Introduction and Learning Objectives

U.S. agricultural cooperatives have a unique historical foundation that affect their legal structure, the state statues under which they are incorporated and even their legal protection to operate. In order to understand the legislation that enabled the creation of U.S. agricultural cooperatives it is necessary to understand market conditions in the early 1900's. Those conditions led to legislation to control monopolies. That legislation led to unintended consequences for farmers and farm organization. That eventually led to legislation (the Capper-Volstead Act [CVA]) that enabled the creation of U.S. agricultural cooperatives. The CVA provided limited immunity from antitrust legislation for specified activities. It also specified the organization form that was required to qualify for that immunity, After reading this chapter you will understand that historical foundation for U.S. agricultural cooperatives and how it led to the enabling legislation, The chapter will also help you understand the legal structure of agricultural cooperatives and the complexity of qualifying for and maintaining immunity from antitrust laws.

Background

The 1800s were a period of rapid growth in the U.S. economy. Rapid industrialization was driven by the expansion of railroads, financed by large banks. The U.S. government initially encouraged the growth of large scale industry but then became concerned over the power of wealthy individuals. Industrialists, often referred to as *robber barons*, took a stranglehold on American industry. They formed *trusts*, which were corporations or combinations of corporations formed mainly for the purpose of regulating the supply and price of products. Some trusts eliminated competition and essentially monopolized industries or geographic markets. The robber barons found that by forming trusts, they could fix prices, make excessive profits, and accumulate wealth. During the 1800s, the number of millionaires in the United States rose from 300 to 4,000. The robber barons had great influence over politicians, which led to increased economic benefits in terms of tariffs and discriminatory railroad rates and rebates.

The Sherman Antitrust Act of 1890

The Sherman Antitrust Act was created to respond to the public outcry against monopolies and their damaging effect on prices, consumers, and small producers, including farmers. As the name implies, the Sherman Antitrust Act sought to rein in formation and activities of trusts created to restrain free competition. The act sought to prohibit anticompetitive practices and the formation of monopolies.

The Sherman Antitrust Act explicitly prohibited some activities, referred to as per se *offenses*. Under the *per se* rule, the defendant is not allowed to offer competitive justifications. All the government has to do is establish that the *per se* offense existed. Traditional *per se* offenses included:

- **Price fixing activities:** Agreements between competitors or between a manufacturer and distributor to establish the price to be charged to a third party. Price fixing can include an element of price such as credit terms, interest rates, or a formula for determining prices.
- **Market division between competitors:** Agreements by businesses to allocate customers and territories. Such agreements allow one firm to raise prices with the knowledge that other firms will not compete for the customers.
- **Certain types of boycotts:** The practice of one or more competitors boycotting or refusing to deal with a third party that refuses to participate in price fixing. However, other less severe boycotts may be judged under the rule of reason, which is explained in the following section.

While Section 1 of the Sherman Antitrust Act deals with *per se* offenses, Section 2 prohibits monopolization or attempts at monopolization, which are judged under the rule of reason. This rule analyzes whether there was an intent or conspiracy to monopolize. Under the rule of reason, the firm must have both monopoly power (the ability to control prices or exclude competition) as well as the willful acquisition to acquire or maintain that power. Market power that developed from the growth or acquisition of a superior product or an historical accident might not be deemed illegal under the rule of reason. High market share may not be sufficient to establish monopolization. For example, if there are no barriers to entry, a firm with high market share could not unfairly raise prices without other firms entering the market and competing.

Section 2 of the Sherman Act also prohibits attempts to monopolize. This prohibition is aimed at conduct by a firm with market power that falls short of a monopoly but that, if continued, could pose a dangerous probability of achieving a monopoly. Again, under the rule of reason, an agreement between firms was judged on both the procompetitive and anticompetitive effects on competition in a defined relevant market.

Although intended to curb the anticompetitive activities of large firms such as trusts, banks, and railroads, the first groups affected by the law were farms and labor associations. If read literally, Section 1 of the Sherman Act prohibits any coordination of supply or pricing among competitors. Agricultural producers are technically competitors in the same industry, so any type of joint marketing activity by farmers could be challenged as illegal price fixing.

The Clayton Act of 1914

In an attempt to address the unintended consequences of the Sherman Antitrust Act, U.S. Congress passed the Clayton Act of 1914. This act provided exemptions from antitrust laws for both associations of producers and labor organizations. The part of the Clayton Act applying to producer organizations provided (U.S. Code, 1914):

> Nothing contained in the antitrust laws shall be construed to forbid the existence and operation agricultural, or horticultural organizations, instituted for the purpose of mutual help, and not having capital stock, or conducted for profit, or to forbid or restrain individual members of such organizations from carrying out the legitimate objects thereof, nor shall such

organizations, or the members thereof, be held or construed to be illegal combinations or conspiracies in restraint of trade, under the antitrust laws.

The Clayton Act therefore exempted "associations for the purpose of mutual help, not having capital stock or conducted for profit" and proved to not be very useful for producer organizations. First, while it would clearly exempt a producer organization focused on education, it was unclear whether it would exempt an organization (e.g., modern-day cooperatives) that was formed to create profits and distribute them to the members. Section 6 of the Clayton Act also did not clearly define the activities in which producer organizations could engage (it merely said "legitimate objects" of such organizations) (U.S. Code, 1914). The phrase *without capital stock* also suggested that producer organizations could not raise investment and fund infrastructure such as grain bins, warehouses, or cotton gins.

The failure of the Clayton Act to effectively exempt producer organizations from antitrust prosecution led Congress to look for a clear definition of the types of organizations and activities they were seeking to protect. Minnesota Representative Andrew Volstead, one of the sponsors of the CVA, argued that corporations were owned by numerous stockholders who, through the corporation, were allowed to act collectively. He argued that the flaw in interpretation of the Sherman Antitrust Act was the assumption that each farmer was a separate competing business entity. He maintained that it was impractical for farmers to combine their farms into corporations, but when they combined their crops with their neighbor to more efficiently market them, they were charged with anticompetitive behavior. In crafting the legislation to create an exemption for farm organization, Representative Volstead and the other sponsors looked to the concept of cooperatives and Rochdale principles.

The Capper-Volstead Act of 1922

The CVA has been termed the magna carta *of agricultural cooperatives* in the United States. The law provided (1) an explicit exemption from antitrust legislation for producer organizations, (2) specified the required structure of an organization to qualify for the exemption, and (3) described the protected activities that such organization could engage in.

Exemption and Description of Protected Activities

The CVA states (USDA, 1995):

> Persons engaged in the production of agricultural products as farmers, planters, ranchmen, dairymen, nut or fruit growers may act together in associations, corporate or otherwise, with or without capital stock, in collectively processing, preparing for market, handling, and marketing in interstate and foreign commerce, such products of persons so engaged. Such associations may have marketing agencies in common; and such associations and their members may make the necessary contracts and agreements to effect such purposes.

In simple terms, the CVA allowed producers to do everything necessary to collectively market their crops and therefore provided marketing cooperative with an exemption from antitrust prosecution. The act further stated that the U.S. secretary of agriculture has the enforcement authority to challenge the operation of an association if it has unduly enhanced prices. For those two reasons, the CVA is said to provide limited immunity to antitrust prosecution.

Required Structure of Organizations

The CVA continued by stating (USDA, 1995):

> Provided, however, that such associations are operated for the mutual benefit of the members thereof, as such producers, and conform to one or both of the following requirements:
>
> First. That no member of the association is allowed more than one vote because of the amount of stock or membership capital he may own therein, or,
>
> Second. That the association does not pay dividends on stock or membership capital in excess of eight percent per annum. And, in any case, to the following:
>
> Third. That the association shall not deal in the products of nonmembers to an amount greater in value than such as are handled by it for members.

Therefore, (considering that the CVA required eligible associations to be owned by agricultural producers) the act specified four restrictions on associations. The association (cooperative) must be owned by agricultural producer, must operate for mutual benefit (distribute profits as patronage), and must either use one-member-one-vote or limit dividends to 8%. The CVA requirements were integrated into many states' cooperative statutes. In many cases, the states did not note the *or* requirement and included both the voting and dividend restriction in their statues.

Qualifying and Retaining Capper-Volstead Status

The CVA states that the immunity is available one to persons engaged in the production of agricultural products as farmers, ranch owners, dairy owners, and nut or fruit growers. Courts have held that only producers may be members if the cooperative wants to maintain CVA protection. What is more, the presence of even one producer-member disqualifies the cooperative from retaining its CVA status. That has been termed the *not even one rule.*

This principle was first articulated in *Case-Swayne, Inc., v. Sunkist Growers, Inc.* (1967). Sunkist, as a federated cooperative, contained both local cooperatives and growers as members. More than 80% of its members were growers—natural persons, corporate farms, and partnerships that were engaged in growing fruit. Five percent of its members were fruit growers who also owned packing facilities. The remaining 15% of Sunkist members were packing houses that contracted with Sunkist growers. The U.S. Supreme Court denied Sunkist CVA status because it had 15% nonproducer packinghouses as members.

In a separate case involving the National Broiler Marketing Association (1978), the Supreme Court denied CVA status to an association of poultry producers because some of the producer-members did not own flocks but merely contracted with producers using risk-sharing contracts. The court decision therefore specified that in order to be classified as a producer, an individual had to take on all the risks of production. While the case involved integrated production, the courts did not comment on whether a producer who maintained their poultry production operation but also owned and operated a processing plant that contracted additional volume would be an eligible member. As farms have become more complex and forwardly integrated, this question has arisen more frequently. While it is clear that an individual must take on all production risk to be a producer, it is not clear whether a producer who simultaneously operates a processing operation is still an eligible member.

Recent Challenges to Capper-Volstead

Perhaps the biggest question raised in recent lawsuits is whether preproduction supply-management activities—such as jointly agreeing not to plant crops or to raise fewer animals—is a protected "marketing" activity under the CVA. The resurgence of this issue is surprising, since prior courts have already found that protected marketing under the CVA includes direct price setting and the actual destruction of products to remove them from the marketplace. If a farmer can legally destroy products, why cannot they simply decline to produce them in the first place? However, in its advisory opinion, the Idaho potato court found that the CVA does not protect such preproduction supply-management activities. The court noted that no prior courts had explicitly approved these activities, the CVA itself does not expressly endorse them, and a 1977 Federal Trade Commission statement indicated that they were not protected. Additionally, the court noted that farmers had an incentive to increase production if prices rose because of a cooperative's activities but that this incentive is missing if preproduction supply-management activities are permitted and future production is, in effect, shut off. Accordingly, the court concluded that preproduction supply-management activities are not protected under the CVA.

Whether the court will continue to adhere to the distinction between postproduction and reproduction supply-management activities remains to be seen. Similar arguments have been made by plaintiffs in a Pennsylvania eggs case and a California litigation involving the National Milk Producers Federation's herd retirement program. There was also an effort during the Obama administration to reexamine whether the CVA is still needed. Clearly, the political environment toward the CVA and cooperatives is not as strong as it was when the act was enacted.

Summary

Three laws affected the structure of U.S. cooperatives. The Sherman Antitrust Act of 1890 prohibited a wide range of anticompetitive behaviors. While intended to curb the activities of trusts, railroads, and banks, farm associations became the unintended target of the Sherman Act. Congress attempted to address the problem through the Clayton Act of 1914. This act proved to be ineffective since it applied only to organizations conducted not for mutual benefit and without capital stock. The Clayton Act did not therefore appear to apply to farm associations investing in marketing and storage infrastructure. It also did not explicitly describe the permitted activities of a farm association.

Drawing on the Rochdale principles, the CVA became the *magna carta* of U.S. agricultural cooperatives, providing them with limited exemption from antitrust prosecution and detailing organizational restrictions as well as permitted activities. The CVA applies only to marketing activities and therefore marketing cooperatives. It also has strict provisions that every member must be a qualifying agricultural producer who bears the entire risk of the production. Recent challenges to the CVA have centered on the question of whether a cooperative could engage in preproduction supply control in an effort to curb excess production and enhance prices.

References

U.S. Code. (1914). *The Clayton Act* (15 U.S.C. § 12 et seq.). Act of October 15, 1914, ch. 323, 38 Stat. 730.

United States Department of Agriculture. (1995). *Understanding Capper-Volstead* (USDA Rural Business and Cooperative Development Service, Coop Information Report No. 35). U.S. Department of Agriculture.

CHAPTER 9

Managing a Cooperative

Introduction and Learning Objectives

As in any business organization, the manager has a major influence on the success of a cooperative business. Management of a cooperative, or any organization, is both an art and a science. The basic functions of management provide some insights into the management process. In addition to the challenges of managing any organization, managing a cooperative requires unique skills. Understanding those management challenges provides additional insight into the nature of a cooperative business. Some of the challenges of managing a cooperative relate to the complexities in how members can benefit from participating in a cooperative. After reading this chapter, you should have a better understanding of the unique aspects of managing a cooperative.

Management Defined

There are many alternative definitions of *management* (Koontz & O'Donnell, 1959). One is that it is "the process of designing and maintaining an environment in which individuals can work with others efficiently and effectively to accomplish set goals." Other definitions of *management* include phrases such as "interacting with people to achieve results," "using resources to achieve objectives," or "balancing tradeoffs." Common themes across the various definitions of the term *management* are the concept of allocating scarce resources and directing individual efforts to achieve the organization's purpose. Fully describing the functions and processes of business management is beyond the scope of this chapter. Instead, our major focus is to identify and discuss how the unique nature of the cooperative firm affects its management.

Management of a cooperative is both an art and a science. Some aspects of management relating to operations and finance are very structured. As an example, the process of budgeting, which is part of the planning process, can be broken down into component steps and analyzed. On the other hand, many aspects of management involve working with human resources, communicating within and outside the firm, analyzing competitors' actions, and setting strategy. Those processes are more complex and involve both an art and a science.

The manager of a cooperative firm has many roles. The manager is the figurehead and spokesperson for the cooperative, representing the firm to external stakeholders. The manager is also a leader, guiding the activities of the employee group, and is also part of the overall leadership in tandem with the board of directors. A cooperative manager is the gatekeeper or liaison between the board of directors and the employee group. The manager communicates the board's strategic goals to the employee group and communicates employee concerns and needs to the board. The cooperative manager both receives and disseminates information. They also address disputes and job performance issues among the employee group, negotiate with suppliers, and make myriad necessary day-to-day decisions. One of the key roles of the manager is to allocate the cooperative's scarce resources in both financial and human capital to best achieve the organization's overall goals of efficiency and profitability. This requires an understanding of operations and finance and ability to take calculated risks and adjust plans as needed. It requires a broad skill set, including financial knowledge, understanding of operations, people skills, and other specific skills, such as financial analysis, project management planning, critical thinking, and decision-making.

Functions of Management

The process of management is often separated into the functions of planning, organizing, staffing, directing, and controlling. The related concept of coordination is also sometimes added along with the concept of meeting social and ethical responsibilities. All these functions are present in the management of a cooperative. While the board of directors has the primary responsibility for strategic planning, the manager typically works in tandem with the board, suggesting strategies and analyzing alternatives. This can create unique challenges for the cooperative manager. Although strategic planning is primarily the board's responsibility, board members—particularly newly elected members—may not have a good understanding of the industry environment facing the cooperative. The manager is responsible for short-range planning, including budgeting, and setting the daily, weekly, and seasonal priorities. In most cooperatives, the board of directors establishes a long-range plan and overall budget for fixed asset replacement, and the manager works within those parameters to make specific decisions on equipment replacement. The cooperative manager must work to maintain the separation between their responsibilities in short-term planning and the board's role in long-run and strategic planning. Because they are customers and operationally focused, some board members want to become overly involved in equipment-related decisions. In addition to routine equipment replacement, most cooperatives have one or more ongoing infrastructure replacement or expansion projects. Overseeing these is an important aspect of the manager's planning role.

The *organizing function* refers to the manager's efforts to direct the cooperative's processes and people. A cooperative manager may decide to organize activities around geographic locations with a mid-level manager at each branch location or create functional departments with managers overseeing, grain, agronomy, fuel, and other functions over the entire cooperative. The manager also orders and positions fertilizer and other farm supplies, decides where to hold grain inventory. and develops plans for harvest and peak demand periods. All these activities involve organizing the cooperative's physical and human capital assets.

The staffing function is closely related and involves recruiting, hiring and onboarding employees, as well as evaluating their performance, setting compensation and incentives, and developing their skill sets. I often describe the staffing function as "Getting the right people on the bus, the wrong people off the bus, and then getting the people on the bus is the right seats!"

As the name implies, the *directing function* involves instructing and guiding the employee group. In terms of the previous analogy, directing involves telling the person in each seat on the bus what to do. It involves straightforward activities, such as assigning employees to specific tasks, as well as more subtle actions, such as providing encouragement and feedback to guide and improve job performance. In a cooperative where most customers are also member-owners, employee interactions with customers are critical. Part of the directing role of the cooperative manager may be to improve the customer relationship skills of front-line employees.

The controlling function of management is the process of checking actual performance and outcomes against the plans and goals. In general, the controlling function involves establishing some sort of standards, measuring performance against the standards, and making corrections when there are deviations. The controlling function can also involve ensuring that the cooperative is in compliance with its own internal policies and all external laws and regulation. In a cooperative, efficiency is extremely important. Pricing affects the cooperative's profits as well as the members at the farm level. For that reason, efforts to improve a cooperative's profitability often focus on improving efficiency. A cooperative manager may monitor the cost per acre of fertilizer application or the per-bushel cost of grain management to ensure that they are within historical standards or industry benchmarks. Other aspects of control involving monitoring the quality and quantity of both commodity and input inventories.

The Unique Aspects of Managing a Cooperative

A cooperative manager needs to operate an efficient and effective business organization, respond to their competitive environment, manage human resources, meet customer needs, and comply with all applicable rules and regulations. Those challenges are obviously not unique to the cooperative business model. There are some area in which the cooperative managers face different challenges from those of managers of other types of firms and therefore need different skill sets. Some of these differences related to the frequency and nature of contact with the owners, the complexity of the cooperative objective function and the avenues by which members receive benefits, and the balancing act between protecting the viability of the cooperative and maximizing returns.

Closer, Longer-Term Relationships With Customers

In most firms, the manager has relatively infrequent contact with the owners. The manager of a publicly held corporation may report to shareholders on a quarterly or annual basis, with the actual contact limited to the select shareholders who have major ownership holdings. Depending on the size of the firm, the manager may or may not have interaction with customers. If a customer in an investor-owned firm (IOF) becomes dissatisfied, they may discontinue their use of the firm. They are unlikely to have future contact with the manager. In a cooperative, the manager has daily contact with the cooperative's members, who are also customers and owners. Member-owners have frequent contact with the manager and value the opportunity to express their preferences and opinions regarding the cooperative's operations. A dissatisfied member still has investment in the cooperative and will remain an owner for many years until the equity is revolved. For these reasons, the cooperative manager has a closer and longer-term relationship with customers and owners.

As mentioned, the manager of an IOF must respond to the needs and opinions of the major shareholders. Because most cooperatives operate according to a one-member-one-vote structure, the cooperative manager is under greater pressure to communicate and respond to every member. When it comes to a vote on a merger or restructuring or even a vote on a bylaw amendment, the smallest producer has the same weight as that of the largest customer.

Cooperative Managers Work in a Spotlight

Because of their frequent interactions with member-owners, the cooperative manager tends to work in a spotlight. Members judge the performance of the cooperative and manager during every transaction. Customer-owners are used to expressing their opinions directly to the manager. In single-location cooperatives, many members expect to be able to drop into the manager's office on every visit. Additionally, cooperative managers are more likely to have a hands-on role in operations relative to their counterpart in IOFs. All these factors can make decisions on product line, inventory levels, hours of operation, or seasonal locations more political relative to other business forms.

Cooperative Managers Have Different Controls and Incentives

In an IOF, the manager focuses on meeting the objectives of the major stockholders or owners. These aims are generally to maximize the profits of the firm. In order to align the manager with those objectives, many managers of IOFs have compensation or bonuses tied to the firm's profitability. In a publicly owned firm, the manager may receive stock options that become valuable if the profits increase and the stock prices rises. In a cooperative, equity is held only by users and doesn't appreciate in value. This eliminates the possibility of using stock options as an incentive. Because the cooperative's profits is based on transactions with member-users, profitability is not the only goal. This makes it more difficult to base compensation and incentives on performance. Producer-members may also object to the cooperative diverting returns from patronage toward employee and manager compensation. A cooperative manager must also respond to every member (who often has an equal vote). This makes their job more complex relative to the manager of an IOF, who can concentrate on the concerns of the majority stockholder.

More Complex Objective Function

The objective of an IOF is to maximize the return on investment of the owners. In the cooperative business model, the objective is more complex. Members also have transactions with the cooperative, purchasing inputs and selling commodities. The prices of those transactions affect their returns at the farm level. The cooperative's role in providing services and creating a stable market or source of supply also affects their farming profitability.

Distributing Returns Is More Complex

In an IOF, profits may be distributed as dividends in proportion to ownership. Many publicly traded firms do not distribute profits but rather retain the funds as retained earnings. The firm's stock price is based on investor projections of future profits, so existing members benefit from the stock price appreciation. In a cooperative, the distribution of returns is much more multifaceted. Cooperative distribute funds as cash patronage. They also distribute patronage in the form of qualified and nonqualified stock, which have different tax treatments. When the cooperative distributes patronage in the form of equity, they are both distributing something of future value (revolving equity) and also retaining funds. Retained patronage is simultaneously a method of retaining funds and distributing benefit. Cooperatives can also retain profits as unallocated equity, also called *unallocated retained earnings*, which is a general reserve fund. All the distribution decisions have impacts on the cooperative's cash flow, the cooperative's equity level, the individual's equity ownership, and members' return from the cooperative.

Four and a Half Ways Members Benefit From a Cooperative

There are at least four clear avenues of benefit to members from their cooperative. As discussed, the first is through the prices paid or received. Favorable prices provide an immediate benefit, which is popular with all members, particularly those younger who have little invested in the cooperative. These members often say, "Why doesn't the cooperative simply increase prices paid on commodities rather than creating profits that must be distributed?" The problem with offering favorable prices is that the manager may underestimate or overestimate costs and end up operating at a loss. The cooperative model also depends on retained profits to both create member equity and finance infrastructure reinvestment and equity retirement. The members of a cooperative with favorable prices may also not appreciate the performance of the cooperative since its profits will be low. Additionally, the cooperative's competitors may meet the favorable price, resulting in the cooperative benefiting nonmembers as much as it does members.

One of the major avenues of cooperative member benefit is through patronage, which may be in a combination of cash and stock (retained patronage). The structure of patronage creates a balancing act. The cooperative must balance cash patronage, which provides an immediate return to the member with retained patronage, which provides cash flow for infrastructure reinvestment and equity retirement. In addition to the balancing act between member returns and the cooperative's cash flow, patronage refunds create a balancing act between member groups. Younger members may favor retaining funds for infrastructure reinvestment, whereas older and inactive members will push the cooperative to concentrate on equity retirement.

The topic of infrastructure reinvestment suggests the next major avenue for cooperative member benefit: through access to the firm's services. In some cases, cooperatives offer services and market access that would otherwise be unavailable to the members. For example, if sugar beet producers did not have access to a processing cooperative, they would be forced to grow less profitable crops, such as corn or soybeans. In other cases, the cooperative may provide superior services. In the case of a large-scale grain farmer, the cooperative's ability to rapidly dump trucks and handle grain is a major value component. If the grain farmer is forced to wait in a long line at the dump pit, they may end up having to idle combines and decrease their efficiency on the farm.

While it is not common for a cooperative to pay dividends on invested capital, this is another potential avenue of benefit. Depending on the state of incorporation and the section of the tax code under which the cooperative is operating, the dividends may be restricted to 8%. That restriction, often referred to as the *subordination of capital*, goes back to the original Rochdale principles. It should be noted that the Rochdale pioneers operated in an environment of low interest rates. One could argue that the intent of the principles was that there should be only a reasonable return on invested capital so that the majority of profits could be distributed as patronage. In the United States, most agricultural cooperative have elected not to pay dividends on invested capital. In recent years, a few cooperatives have experimented with revolving equity into a dividend-paying class of preferred stock. While this requires diverting some profits into dividend payments, it allows the cooperative to maintain the equity base. Many members appear satisfied by a longer revolving period if they receive dividends at some point. The new hybrid cooperative-IOF business form obviously provides a return on invested capital for the investor class of members.

The final avenue of benefit, which is sometimes referred to as *half of a benefit* since it accrues to both members and nonmembers, is the impact of the cooperative on the competitive environment. This brings up the concept of the cooperative as a ***competitive yardstick***. The competitive yardstick is a well-known analogy for justifying why the presence of a cooperative is thought to promote competition. The concept is that because the cooperative distributes profits to the users in proportion to use, the firm has no incentive to charge excessive prices. The cooperative therefore provides a comparison, or yardstick, of the fair and appropriate prices in a marketplace. The cooperative therefore helps keep other firms competitive and, in doing so, creates a competitive marketplace. This is often referred to as the ***invisible benefit of the cooperative***.

Member Control in a Cooperative

Democratic control is a fundamental cooperative principle. However, the processes by which members exercise their democratic control are often misunderstood. Cooperative members generally do not vote on policies, nor do they vote on hiring employees or even the chief executive officer (CEO). Members vote to elect board members, with the vote generally occurring during the annual meeting. During some annual meetings, they may be asked to vote on amendments to the bylaws or articles of incorporation. They may also vote on a major structural changes such as merger or dissolution, during a regular annual meeting or special meeting of the membership. While it is important, democratic control through voting is limited.

Members also exercise control by voicing their opinions to the board and manager. The frequent contact with members is one of the factors that makes managing a cooperative unique. Because every member has the same voting right in most cooperatives, the manager must respond to every member, not only the major users or equity holders. The final avenue of member control is for the member to vote with their feet and discontinue using the cooperative. However, because of their equity holdings, inactive members may still choose to try to influence the cooperative's decisions.

Due to the greater interaction between the board and CEO and the membership, decisions in a cooperative tend to become more political. Decisions on making a location seasonal or changing the product line or hours of operations can become major issues with the membership. This leads to what has been termed the "***cooperative dilemma***: Do we operate as a membership organization or as a business enterprise?" (Kenkel, 2019). Most would argue that the manager should operate an efficient business organization, but they must also realize the inherent differences that arise when the customers are also owners who expect to have frequent opportunities to provide input to the manager.

Summary

Management of any firm is a complex process that can be considered both an art and a science. Managing a cooperative has some unique aspects. The cooperative manager has frequent and long-term contact with the owners. A cooperative manager tends to work in the spotlight, with every operational decision scrutinized by the member-owners. The cooperative manager operates under a more complex objective function because members benefit both at the farm level, from use, and at the cooperative level, from profits. Distributing profits is complex because the manager must balance the cash needs of the cooperative with the members' preference for cash. Members have various streams of benefits from a cooperative, including cash and revolving equity patronage, access to services, and infrastructure. The cooperative may also be providing a service through its mere existence. All these avenues of benefits must be balanced. Members also frequently express opinions to the cooperative manager, which is one of their major avenues of control.

References

Kenkel, P. (2019). *Managing a cooperative.* Cooperatives Community of Practice. https://cooperatives.extension.org/managing-a-cooperative/.
Koontz, H., & O'Donnell, C. (1959). *Principles of management* (Vol. 12). McGraw-Hill.

CHAPTER 10

The Cooperative Board of Directors

Introduction and Learning Objectives

Every individual should understand the role of the board of directors. While there are differences across the business forms, the board of directors is an important part of the governance of investor-owned corporations, cooperative corporations, and not-for-profit corporations. Understanding the functions and role of the board of directors will help you better understand those specific business forms. Additionally, many individuals are asked to run for or serve on a board of directors sometime in their career. Any prospective board member should understand the important legal and fiduciary responsibilities as well as the commitment of time, skill, and resources. While there are many unique aspects of the cooperative board of directors, understanding the role and responsibilities of the board in a cooperative firm also provides insights into the role of directors in other firms and not-for-profit entities.

The Cooperative Board of Directors

The cooperative board of directors has the primary legal authority for the firm. Except for a few major decisions, which are specified in the articles of incorporation or bylaws to require a vote of the membership, the board has top-level decision control to oversee the corporation and ratify important decisions. In practice, the board delegates much of the authority for operational decisions to the manager. In an effectively governed cooperative, the board oversees the manager and makes strategic and policy decisions, while the manager supervises employees and makes operational decisions. The board hires, directs, and evaluates the general manager or chief executive officer (CEO). We typically say that the board has one employee—the CEO, who manages all the other employees. The board works with the CEO to develop and update the cooperative's mission and set strategic goals. The board makes long-run decisions on capital structure. It is the board of directors who signs a long-term loan agreement. The board also makes annual decisions in distributing profits in cash and retained patronage and whether to retire previously issued equity. Through these decisions, they determine the cooperative's level of equity. The board also approves legal, accounting, insurance, and banking relationships and oversees the annual audit. It also establishes policies. Many students assume that cooperative members make policies, but that is the board's role. Most cooperatives have hundreds of complex policies, and it would be impractical to debate all of them during an annual meeting. The board chair presides over the cooperative's annual membership meeting.

As stated, the board of directors is the governing body of the cooperative and has wide-ranging authority as specified in the articles of incorporation and bylaws. A board member has authority only when acting within a duly authorized board meeting. At any other time, a board member has no more authority or stature than any other cooperative member. The only exception to that rule would be when the board appoints a subcommittee of the board to undertake a particular task. A subcommittee would have authority to conduct investigation or other activities outside of the board meeting, provided that it pertained to the committee's specific charge. Board members have a responsibility to communicate cooperative issues to the membership and convey members' concerns to the board. That communication role, while important, does not imply that board members have any individual authority to make decisions or promises.

Board members are elected by the membership to oversee the cooperative's activities and safeguard the members' investments. Because of these expectations and the board's role as the top-level governing body, serving on the board of directors comes with legal responsibilities. From a legal perspective, the board of directors has three fundamental duties.

Duty of care: This involves taking the care and exercising the judgment that any reasonable and prudent person would exhibit in the process of making informed decisions, including acting in good faith consistent with what you as a member of the board truly believe is in the best interest of the organization. The law recognizes and accepts that board members may not always be correct in their choices or decisions, but it holds them accountable for being attentive, diligent, and thoughtful in considering and acting on a policy, course of action, or other decision. Active preparation for and participation in board meetings where important decisions are to be made is an integral element of the duty of care. A board that did ask for a feasibility study prior to making a major decision, or a board member who did not attend meetings or did not review the meeting packet prior to making a decision, might be viewed as violating the duty of care.

Duty of loyalty: This calls on the board and its members to consider and act in good faith to advance the organization's interests. In other words, board members will not authorize or engage in transactions except those in which the best possible outcomes or terms for the organization can be achieved. This standard constrains a board member from participating in board discussions and decisions when they as an individual have a conflict of interest (i.e., their personal interests conflict with organizational interests, or they serve multiple organizations whose interests conflict). A board member who participated in a discussion about the cooperative's purchase of a tract of land that they or their family members owned would likely be viewed as violating the duty of loyalty. When a possible conflict of interest arises, the board member should inform the board of the possible conflict, and if the board concurs that it is a conflict of interest, the board member should recluse themselves from the discussions. The minutes would reflect that the board member left the room during the discussion and voting and returned after the decision was reached.

Duty of obedience: This requires obedience to the organization's mission, bylaws, and policies as well as honoring the terms and conditions of other standards of appropriate behavior, such as laws, rules, and regulations. Many cooperatives have a code of conduct for directors, which includes items such as regarding the cooperatives' credit policy. A director who does not keep their account with the cooperative current might find themselves in violation of the code of conduct and in danger of violating the duty of obedience.

The 'Fiduciary Responsibility' of Boards

Boards and board members often are reminded that they have a "fiduciary responsibility" to the cooperative and, ultimately, the membership. At its core, this is the responsibility to treat the resources of the organization as a trust, and the responsible board will ensure that these resources are utilized in a reasonable, appropriate, and legally accountable manner. While the phrase *fiduciary responsibility* often is used to refer especially to financial resources, it applies to the stewardship of all the assets and resources of the organization. In general, the appropriate exercise of fiduciary responsibility includes developing and implementing an ongoing system to monitor and assess the cooperative's financial condition and performance; establishing a system for budgeting the cooperative's financial resources; adopting a set of policies to govern the acquisition and use of financial resources, including a formal risk management plan; and implementing a formal external review process, such as an independent audit to assess the cooperative's fiscal condition and performance, including the effectiveness of its internal controls.

In addition to those responsibilities, the board has general responsibilities to both the membership and the manager. The board obviously has the responsibility of making decisions that are in the long-run interest of the cooperative. In that process, the board member has the responsibility to convey member concerns to the board and manager. While a board member may have elected to represent a particular geographic district or type of producer, their duty is to present the concerns of their district to the board as a whole. However, when making decisions, their primary concern must be whether the action is in the best interest for the cooperative. Board members have a duty to keep sensitive information confidential. At the same time, they should objectively communicate nonconfidential information to members. A board member may also need to explain the logic and rationale for a decision made by the board. While it is important for a board to actively debate issues and provide dissenting opinions, when a decision is reached, the board member should support the majority decision and provide a unified voice to the membership. Cooperative members often view board members as their elected representative and expect them to reflect their personal views regarding a decision. While the board members are elected by the membership, their primary responsibility is to protect the cooperative's viability. In the long run, that responsibility aligns with every member's interest, but in some decisions, protecting the cooperative may come as a short-run sacrifice by members.

The cooperative board also has responsibilities to the manager and/or CEO. The board should work with the manager in developing strategic plans and visions that will help guide operational decisions. The board should also spell out goals for the manager, objectively evaluate their performance, and actively manage their compensation and incentives. The director's most important responsibility with the board is develop a good teamwork relationship. In performing as a team, it is essential that the board understand their role and not intrude into employee and operational issues. Because of their background in running their own farming operations, many board members tend to be operationally focused. They often unintentionally shift away from their oversight and strategic role and move into second-guessing management decisions.

The Business Judgement Rule

The business judgement rule is a doctrine that developed out of case law and generally frees board of directors of a corporation (including a cooperative corporation) from personal liability from business decisions. The term comes from a Delaware case in which the court stated, "We will not substitute our notion of what is or is not sound business judgement if the directors of the corporation acted on an informed basis, in good faith, and [with] an honest belief that the action taken was in the best interest of the company(Irwin H, 2007)."

In simple terms, the business judgement rule says that a board member will not suffer liability from a bad decision as long as the director was careful and informed and their intent was for the corporation's best interest. While the business judgement rule is a distinct concept from the duties of care, loyalty, and obedience, it is easy to see similarities in the concepts.

Common situations in which a director may lose protection under the business judgement rule include a conflict of interest and lack of due diligence. A conflict of interest could occur if the director is a party to a transaction, stands to benefit personally from the decision, or has a financial interest that would be reasonably expected to influence their judgement. Another common situation resulting in a loss of business judgement rule protection is lack of care. If a board member or the board as a whole does not exercise the care and due diligence that a prudent person would use in a similar situation, they may not qualify for protection by the business judgement rule. Although some board members think that they are avoiding liability by not reviewing information, they are actually increasing their potential for personal liability by not being fully informed.

Board Structure

Board structure varies across cooperatives. Most cooperative statutes require a minimum number of directors, with five being the most common minimum. If one were to poll cooperative board members, most would likely say that a board size of seven to 10 members would be ideal. As a board's size increases, the logistics of board meetings become more complex, and the amount of discussion time is increased. On the other hand, a larger board has greater ability to create subcommittees to work on specific issues. Cooperatives tend to have an odd number of members on the board to avoid the possibility of a tie vote. Cooperative boards tend to try to work toward a consensus opinion. The board chair traditionally does not vote unless their vote is needed to break a tie. Therefore, the odd-numbered board size does not completely solve the problem of tie votes unless the chair is comfortable with being the tie-breaking vote. Larger cooperatives and those operating over larger geographic regions tend to have larger boards. A board's size also tends to increase after a merger. Many merged cooperatives begin with the members of both former organizations on the board of the merge firm and implement a plan to decrease the board size over time.

Most cooperative elect directors for three-year terms, although shorter and longer terms do exist. A cooperative generally starts with staggered board terms so that only a portion of the directors are up for reelection each year. Some cooperative impose term limits, with three or four terms being a common maximum. Other cooperatives have age limits beyond which a director cannot be nominated for election. These structures create an automatic rotation that some view as creating a healthy turnover on the board. On the other hand, term limits and mandatory retirement can force experienced directors off the board. In many cases, it takes a new director one to two years to get up to speed on the cooperative's operations and activities. Excessive turnover reduces the board's effectiveness. Some cooperatives use nonvoting associate (junior) board members. The associate board structure gives potential board members the opportunity to sit in on meetings and engage in discussions. Although there is no guarantee that the membership will elect a former associate member, these members are often viewed favorably by the membership due to their experience and demonstrated commitment.

Many cooperative boards consider electing nonmember, outside, expert directors. The Farm Credit Act Amendments of 1986 required that all associated associations and banks have at least one nonmember director on their boards. Other cooperatives have had long-term experience with outside directors. Almost all boards that have them find that these directors make a significant contribution. They are typically chosen for expertise in areas member directors do not have, such marketing, finance, plant operations, and management information systems. They are typically elected by member-directors rather than by members. About half of the states make a specific exception to the typical requirement of an all-member board, permitting the appointment of nonmember directors. In these states, the outside directors can have the same powers and rights as other directors, although many statutes typically limit their numbers to one-fifth of the total directors on the board. The tradition of low or only symbolic compensation of cooperative directors can make it difficult to recruit outside directors.

Board Decision-Making

Effective board of director meetings and efficient decision-making is both a science and an art. The board chair usually works with the CEO in determining the meeting agenda. An annual calendar may be used to cover specific topics throughout the year. The board chair has a key role in board effectiveness. In addition to helping set the agenda, the chair leads the discussion and tries to ensure that every board member has a chance to express their perspective. A good board chair encourages board members not to take a stance on an issue until both the advantages and disadvantages have been discussed. When possible, the chair may want to schedule an issue as a discussion item during one meeting and bring it up as an action item for a vote during the next meeting. Doing so helps prevent the board from making rash decisions due to the passions of the moment.

It is likely and even desirable that the board will have disagreement on a decision. A good board has diversity not only in composition but also in their thought process. However, after ample debate, the board must reach a decision. Once a decision is made, the board has spoken, and the board members should respect the majority decision. The board should expect the CEO and membership to respect a 5–4 vote as much as a 9–0 vote. Board members should support the decision of the board outside of the boardroom and attempt to speak with one voice. Even when a board member disagrees with the majority decision, they should avoid sabotaging the board by suggesting to members that the decision was incorrect. Because of the value of acting as a unified board, many boards try to reach unanimous decisions whenever possible. As one board member remarked, "While it is important the we explore and discuss an issue from all sides, we try and reach a consensus opinion. If we can't come to a consensus on an important decision, that might be an indication that we haven't talked through it enough" (Reynolds, 2008).

There are times when a director may want to record a dissenting vote even when it is clear that doing so will not change the majority decision. A board member may have a strong opinion on an issue and hope that a dissenting vote will encourage the other members to carefully consider their decision. In the rare instances when a director feels that the board is operating illegally or unethically, they should cast a dissenting vote and have it reflected in the minutes. In some cases, a dissenting director is not held liable for legal issues subsequent from a majority decision that results in an illegal action. The courts created this structure with the hope that a dissenting director could help prevent a corporation from taking inappropriate actions. It should be emphasized that instances of inappropriate or illegal board decisions are very rare in cooperative firms. The more practical consideration is whether a director feels so strongly about an issue that they are willing to risk creating disharmony within the board.

Board Compensation

As the amount of time and number of responsibilities required of directors have increased, there has been a tendency to raise director compensation. However, there are still many small cooperatives that do not pay any remuneration. In almost all cases, directors are reimbursed for ,expenses associated with their cooperative duties. Historically, cooperatives have merely paid a per diem for each day of cooperative business. For larger cooperatives, this generally varies between $150 and $300 per day (Weick et al., 1997). However, there is an increasing tendency to also pay an annual retainer for time a director spends on cooperative business that is difficult to specify, such as communicating with members, keeping abreast of industry issues, and preparing for board meetings. The amount of the annual retainer varies. In almost all cooperatives, directors receive less compensation than do IOF directors for similar responsibilities.

Board Member Recruitment

Despite the importance of the board of directors, many cooperatives do not invest enough effort in identifying and recruiting potential members. Board member recruitment is complicated by the fact that the selection of the board of directors is a member responsibility. This suggest that the directors should have a hands-off approach to the recruitment and nomination process. Unfortunately, most members have a poor understanding of the operations of a board of directors or how well directors perform. Members may nominate or reelect individuals with good political skills sometimes to the detriment of those with strong business skills. Ideally, both the current board and the CEO should be involved in identifying potential board members while leaving the specifics of the process to the nomination committee. As a cooperative evolves and grows, it may need a director with specific financial, operational, or even political expertise. The fact that cooperative boards are restricted to cooperative members limits the opportunity to recruit board members with specific skills. The more limited pool of potential members makes it essential that the cooperative work to identify and encourage nominating ,the best possible candidates.

Most cooperatives use a nominating committee which is often composed of formal board members. The charge of the nominating committee is to identify a pool of qualified candidates for each open position. This is often difficult both because nominating committees often are not active enough and producers may be unwilling to run against another farmer or against an incumbent director. There are alternative systems for informing the membership about potential board candidates. Some cooperatives merely list the candidates during the annual meeting. Others have potential candidates introduce themselves and make a few comments. Some cooperative allow members to nominate themselves or other members by mail or online. They then circulate information about the slate of candidates prior to the annual meeting. Some progressive cooperatives even conduct voting online prior to the annual meeting and simply announce the new board members during the annual event.

Smaller cooperatives tend to elect directors at large, whereas larger and more complex organizations are more likely to have geographic districts. Some cooperatives have formal or informal systems to maintain balance across specific farming systems or cooperative services. For example, a cooperative might want to keep a balance between irrigated and dryland producers or between grain producers and livestock producers. One of the problems of electing by district or other criteria is that it might not yield the best possible directors. If for some reason a cooperative has several good directors in one district, it may not be logical to elect a less qualified director just to maintain the geographic balance. Directors selected on the basic of district or farming type may also have more tendencies to focus decision-making solely through the lens of that membership group. However, as cooperatives grow, members sometimes feel less connection with them. The district system may help the board stay connected with members across the trade territory.

Board Member Qualifications

In general, there are a few specific qualifications one must meet to be eligible to be nominated for a cooperative's board of directors. Any member of the cooperative who is in compliance with the cooperative's policies and meets the criteria of the board's code of conduct is eligible to be elected. A common clause in the code of conduct is that a board member cannot operate a business in competition with the cooperative. For example, there is no reason why a farmer who also operates a custom fertilizer application business could not be a member. However, it would not be appropriate for them to serve on the board and make decisions that affected the cooperative's fertilizer application activities. A minority of cooperatives require candidates to complete a cooperative board education program prior to being nominated. Most cooperatives do not want to interfere with the members' right to select board members and encourage new board members to attend educational programs.

Although there are relatively few official or legal qualifications for serving on a cooperative board, other than being a member of the cooperative and meeting any qualifications spelled out in the bylaws or policies, there are a great number of desirable qualifications. A board member should be of unquestionable character. They should have the trust of other members, the existing directors, and management. They should also have a realistic vision of the industry and the cooperative. A potential director needs good business judgment. They should be a team player but also have the ability to add independent assessments. Finally, a person considered for a director should have no personal agenda as a director, either for themselves or for a specific subset of members who supported them. That is not to say that the director cannot communicate the perspectives of a membership constituents to the board, but they should base their decisions on what is good for the cooperative as a whole.

Director Education

For the board of directors to play its appropriate role in cooperative decision-making, there is a need to educate and train directors. This is especially true of newly elected directors. A new director is used to considering decisions of the cooperative as a customer. They are now in a different role, with the primary responsibility of protecting the viability of the cooperative. Educational programs can help them understand how to remove their farmer hat and put on their board member hat. Available training sessions cover such topics as the legal responsibilities and liabilities of directors, the distinctions between board and manager responsibilities, understanding financial statements, cooperative equity programs, and components of sound marketing strategies. Most cooperatives conduct their own internal orientation programs to familiarize their directors with internal operations. More formal and Multi cooperative director training programs are offered by state cooperative councils, universities, cooperative banks, and cooperative trade organizations.

Summary

The board of directors plays an important role in many types of organizations. Many individuals will be asked to serve on a board of directors at some time in their life. To do so, it is advantageous to have a basic understanding of the roles and responsibilities of the board. The board's basic duties are the duty of care (taking the care and judgement that a reasonable person would take in a similar situation), the duty of loyalty (working only for the organization and avoiding conflicts of interest), and the duty of obedience (complying with the organization's mission, bylaws, articles, policies, and laws and regulations). Those duties are related to the business judgement rule, which states that a board member will not have personal liability from an errant decision if they acted with due care, in good faith, and without a disability conflict for a valid business reason.

The board of directors sets policy, determines the strategic vision, makes long-term financing agreements, makes annual decisions of patronage and equity retirement, and approves major purchases. Boards must work to become an effective team and develop a partnership with the CEO.

Reference

Reynolds, A. (2008) *Benefiting from the board: A case study*. Staff Paper No. 8. University of Wisconsin Center for Cooperatives. https://resources.uwcc.wisc.edu/CaseStudies/Benefitting_Board_CaseStudy.pdf

CHAPTER 11

Understanding Cooperative Returns

Introduction and Learning Objectives

One of the fundamental characteristics of a cooperative that distinguishes it from other forms of business is that the customers (patrons) are also the owners of the firm. Members are making a joint decision to invest in the cooperative as owners and patronize the cooperative firm. That creates the possibility that the member can benefit at both the farm and cooperative levels. Farm-level returns could include creating more favorable prices, reducing risk through pooling, or obtaining access to services that are otherwise unavailable. The cooperative may also be creating an invisible benefit by keeping the market more competitive. None of those benefits shows up on the cooperative's income statement, and these advantages are separate from the patronage distributed to members. This chapter will help you understand the full dimension of possible returns to the member of a cooperative. It also provides some insights into the decisions and strategies that cooperative leaders should follow as they attempt to maximize member returns.

A Cooperative Can Create Returns at the Farm Level

Both levels of the returns (farm and cooperative levels) are potentially important to a potential member. In theory, a producer would invest to join a cooperative when the perceived returns at the farm level plus the projected returns from the cooperative were higher than alternative investments. For most producers, the relevant alternative investment is investing in their farm operation. So, a producer would be expected to join a cooperative when the benefits from the cooperative at both the farm and cooperative levels were greater than investing in their farming operation.

As you might recall, one of the original Rochdale principles was "goods sold at market prices," and as we will see in later chapters, cooperatives are encouraged to set the prices for both farm supplies and farm commodities at the prevailing market price. In other words, the best practice is typically for the cooperative to price at market levels and concentrate on generating returns at the cooperative level, which can be passed on to the members as patronage. When a cooperative tries to benefit members through favorable prices, it runs the risk of underestimating or overestimating costs and ends up making a loss. The cooperative's competitors also often match favorable prices, so the strategy ends up benefiting nonmembers as much as members. Still, the broader point is that members can potentially benefit at both the farm and cooperative levels, and the cooperative board and manager must consider both effects. This is a unique aspect of the cooperative firm.

There are some instances in which the potential for member-level returns is particularly apparent. Some marketing cooperatives, particularly those marketing specialty fruits and vegetables, operate on a pooling basis. Under that structure, the cooperative does not purchase the commodity but rather gives the member an advance on the final pool distribution. The advance payment serves as a partial payment of the final pooled value of the commodity. Eventually, all the commodities in the pool are sold, and the residual amount is distributed to the members in proportion to volume. One can think of the first payment as relating to the commodity value and the second payment as a profit distribution; however, there is really no clear separation. The pooling cooperative is operating entirely as an extension of the farm. The final value of the commodities combines the value created at the farm level (growing and harvesting the fruit or vegetable) and that created at the cooperative level (processing, packaging, and marketing). However, there is no separation between those values. The advance payment does not necessarily reflect the value of the crop—it is simply an advance on the final projected value of the final product. In a pooling cooperative, all the return comes at the member level. Members may appreciate the importance of their pooling cooperative, but they really have no measure of the value created at the cooperative level.

Another instance where member-level value is important occurs when the existence of the cooperative allows the members to grow a more profitable crop. For example, in the Northern Plains, sugar beet production has historically been more profitable than corn or soybeans. There have been instances when an investor-owned sugar beet–processing facility has decided to close, and farmers have formed a cooperative to purchase the facility or built a processing plant. That raises the question as to why the producers reached a different conclusion regarding the feasibility of a sugar beet–processing plant from that of the investor-owners. It is possible that the producers had a lower hurdle for an acceptable rate of return. What is more likely is that the producers were considering the additional farm-level returns that the processing plant would create by allowing them to grow a more profitable crop. In the case of a sugar beet –processing cooperative, its members can monitor the returns created at the cooperative level and will receive a share of them through patronage. They also receive a farm-level benefit by being able to grow a more profitable crop. The cooperative board and management would need to consider the farm-level returns that were created by the cooperative's existence in addition to the cooperative-level returns.

A more complex example of farm-level returns occurs in farm supply cooperatives when they provide products or services that are important to members but only marginally profitable. A classic example is a cooperative tire shop that fixes flat tires on vehicles and implements. The tire department is frequently not profitable. Members might argue that access to tire repair services is important at the farm level. Cooperative leaders face the dilemma in determining whether the possibility of farm-level returns justifies maintaining a service that is not generating returns at the cooperative level. One challenge is that while members might be opposed to closing the tire shop, they may not consider the farm-level value when they evaluate the cooperative's performance. It is not unusual to see a situation in which members are dissatisfied by their cooperative's profitability while at the same time supporting the cooperative maintaining unprofitable products and services due to the value they provide at the farm level.

While cooperative leaders must consider farm-level returns, whenever possible, the cooperative should try to generate most of the return at the cooperative level. Farm-level returns are difficult to measure and sometimes accrue to nonmembers as well as members. Cooperative boards should scrutinize any investment or activities that are being justified solely on the basis of service differentials, value of existence, or value in reducing risks. Those farm-level returns, although important, are also subjective and may vary from member to member. That does not mean that these returns should be ignored. It means that activities generating only farm-level returns should be examined closely to ensure that such value actually exists and is recognized by the members.

Return on Investment in a Cooperative

In an investor-owned firm, the owners purchase stock or other form of ownership and receive a return through dividends on equity and/or appreciation of the stock value. In an investor-owned firm, there is no return to the users or customers. That makes the calculations of return on investment (ROI) fairly straightforward. The owners' investment and owners' return are readily apparent. As mentioned, cooperative members are both investors and users. That raises some interesting questions regarding measuring a member's ROI and considering what the appropriate return should be. In theory, a member's return from a cooperative includes patronage refunds, dividends paid on equity, and farm-level returns (e.g., price differential, service differential, value of existence, and value of risk pooling). As we have discussed, farm-level returns may be important, but they are difficult to measure. If we focus on cooperative-level returns, we see that members benefit primarily through patronage refunds, which are paid in a combination of cash and revolving equity. The members' returns reflect the cash patronage and present value of the revolving equity, which will be redeemed for cash at a later date. As we will see in later chapters, dividends on invested equity are limited by law, and only a minority of cooperatives pay any dividends on equity capital.

Ideally, a cooperative would provide fair returns to members as both users and owners. Patronage payments (both cash and the eventual return from revolving equity) provide a return to use; however, the return to ownership in a cooperative is indirect. This can lead to a free-rider problem, in which new members receive the benefits from the cooperative with very low investment. It can also provide a disincentive to support long-run investment by the cooperative since the member will benefit only from future patronage. For example, we would not expect older members in a dairy cooperative to support investing to develop branded value-added products. They will likely conclude that they will not be active in the cooperative long enough to receive those benefits through future patronage. They would instead push for the cooperative to increase cash patronage and not retain funds for reinvestment in the cooperative.

Closed cooperatives (new-generation cooperatives) solve the free-rider problem and the issue of incentive for long-term investments by linking investment to usage rights. Members have the opportunity to sell their shares and usage rights to another eligible member with the approval of the board of directors. If the new-generation cooperative is successful, patronage payments will increase, and the usage rights will become more valuable. That structure provides the potential for an appreciation in stock price and an additional aspect of the ROI. In traditional or open-membership cooperatives, the owners' return comes entirely through patronage, which implies that the members must continue to use the cooperative in order to receive a return. Open-membership cooperatives cannot entirely eliminate the free-rider problem but can minimize it by managing equity in a manner that keeps equity investment in proportion to use. Keeping member investment proportional to use helps treat members fairly as owners.

The denominator of the ROI formula (owners' return/equity invested). As we will see in a later chapter on capital structure, the owner's equity in a cooperative business also has some unique characteristics. Some of the most common forms of equity are the direct investment that a member makes in order to join the cooperative, the retained patronage (patronage issued in the form of stock), and the unallocated equity. Unallocated equity is created when the cooperative retains funds without creating equity in individual patron names. Members have a collective claim on the unallocated equity, but it is never redeemed, and the member would receive the value associated with the unallocated equity only if the cooperative was liquidated.

What is the appropriate equity category to use in measuring an owner's ROI? The answer is not entirely straightforward. In a closed cooperative, the vast majority of the equity comes in the form of the direct investment that was made to join and obtain usage rights. That makes it fairly easy to calculate the ROI in a closed cooperative. In open cooperatives, direct investment is only a small portion of total equity, and members make only a nominal investment to join the cooperative. Most of the equity in an open cooperative is created out of the profit stream in the form of allocated revolving equity (stock patronage) and unallocated equity (general reserve fund). Most cooperative scholars agree that the cooperative's ROI should be based on the total member equity. It must be noted, however, that much of that equity may have been earned out of the profit stream, and members may have made very little actual out-of-pocket investment.

At times, cooperative CEOs and boards argue that because the members did not make a direct investment for the majority of their equity, it is not important to measure a cooperative's return on equity. The counter to that argument is that the cooperative is owned and controlled by the members. All these members have alternative uses for funds in their own farming operation. If the cooperative is unable to generate a return on the members' equity, there is an argument for liquidating the cooperative and allowing the members to invest the funds at the farm level. This suggest that one measure for the appropriate ROI in a cooperative is that it should be comparable with returns from noncooperative investment. The long-run return to the stock market has been about 10%. Cooperative boards might consider that level as a goal for return on total equity. Of course, if the cooperative is less risky relative to the typical firm or structured to help the members reduce risk, a lower return might be acceptable.

Guidelines for Creating Returns in a Cooperative

Although this discussion has illustrated the complexities of understanding and measuring the owners' return from a cooperative, there are some general conclusions, as follows:

1. Cooperatives can generate return at both the cooperative level and the farm level. While cooperative leaders must consider farm-level return, whenever possible, the cooperative should try to generate most of the return at the cooperative level. Farm-level returns are difficult to measure and sometimes accrue to nonmembers as well as members. Cooperative boards should scrutinize any investment or activities that are being justified solely on the basis of service differentials, value of existence, or value in reducing risks. Those farm-level returns, while important, are also subjective and may vary from member to member. That does not mean that farm-level returns should be ignored. It means that activities generating only farm-level returns should be examined closely to ensure that such value actually exists and is recognized by the members.
2. In most agricultural cooperatives, the owners' return comes primarily through patronage. That eliminates any direct incentive to hold equity and can make members with short time horizons oppose the cooperative's investment in long-term assets. All members can receive patronage benefits regardless of their investment (the free-rider problem), which can cause members to not support the cooperative's actions in retaining funds and building equity. Closed-membership cooperatives solve this free-rider problem by linking investment with use and requiring substantial up-front investment. Open-membership cooperatives cannot eliminate the free-rider problem but can reduce it by distributing an appropriate portion of profits in the form of equity (which helps members build equity investment) and by managing revolving equity to keep equity in proportion to use.

3. There are some unique issues in discussing the ROI in a cooperative business. The member's return may include farm-level returns, which are difficult to quantify. Additionally, some categories of member equity are created through retained profits and do not represent an out-of-pocket investment. Despite those unique features in measuring and benchmarking ROI, cooperative leaders should strive to make cooperative-level returns equal to or exceeding those from alternative noncooperative investments. Regardless of how the equity was created, members will expect the cooperative to generate a reasonable return on those funds. A cooperative that continually fails to create a reasonable return on equity stands the risk of the members dissolving the cooperative to better deploy the capital in their own farming operations.
4. Cooperative performance cannot be measured solely on the basis of profitability. The members' cooperative-level return will also be affected by how the cooperative distributes profits and the revolving equity time cycle. In addition, there are farm-level returns such as service differentials and risk management. Finally, the cooperative may be generating a return through its very existence by either allowing the producer to grow a more profitable crop or keeping markets competitive. Cooperative leaders must consider these other returns but not use them as an excuse for failing to generate adequate financial returns at the cooperative level.
5. Cooperative leaders should strive to treat members fairly, both as patrons and as owners. The cooperative should allocate a high portion of the profits to patrons and distribute those profits in an appropriate mix of cash and equity patronage. The practice of retaining profits as unallocated equity (which is never revolved to the patron) should be used sparingly. Returning profits in proportion to use through patronage treats members fairly as users. In setting the appropriate mix of cash and equity patronage, the cooperative should consider the cash needs of the cooperative as well as the need for patrons to build equity ownership. The board should attempt to manage equity revolvement in order to keep equity ownership in proportion to patronage. Keeping member investment proportional to use helps treat members fairly as owners.

Summary

Before we focus on measuring financial returns in a cooperative, it is useful to consider the full dimension of returns. Understanding and measuring a cooperative's returns is complex. Members make a joint decision to invest in and patronize a cooperative. That means that they are both users and owners. The cooperative business model is focused on providing returns to use, but the board and manager must strive to ensure that members are adequately invested in the cooperative. Most cooperatives do not provide a direct return on invested capital, but by keeping investment in proportion to use, they indirectly create an ROI and treat members fairly both as users and owners.

Cooperative leaders should also strive for the cooperative to generate an adequate ROI for the members. One benchmark for an adequate cooperative-level return would be a return comparable with the long-run return on the stock market or returns from other investment with a similar risk level as the cooperative. Cooperatives can also generate a return at the farm or user level, which can include price differentials, service differentials, risk reduction, and a value of existence. Cooperative leaders must carefully weigh the impact of those returns; however, to the extent possible, they should work to maximize cooperative-level returns. In the long run, members will choose to remain a member of a cooperative as long as combined returns at the farm and cooperative levels exceed that of any alternative use of their equity capital.

CHAPTER 12

Profit Distribution and Cash Flow

Introduction and Learning Objectives

In the introductory chapter, we introduced the topic of patronage and discussed the mechanics. In this chapter, we take a closer look at a cooperative's choices for distributing and retaining profits. Patronage allocations are part of those decisions, but patronage is only one of a portfolio of alternatives for the use of cooperative profits. Distributing and retaining profits has major implications for the financial return of the members and the cooperative's financial stability. These decisions also affect the cooperative's balance sheet and thus the growth of the cooperative and its ability to reinvest in the productive assets that create member benefit. In making decisions, the cooperative board must perform a balancing act, considering both the members' need for return and the cooperative's requirement for cash flow. The choices are unique to the cooperative business model. This chapter will help you unravel and understand the complex alternative uses for a cooperative's profits.

The Critical Financial Decisions of Profit Distribution and Retention

Cooperatives invest in long-term assets, such as buildings and equipment, as well as short-term assets, such as inventory and accounts receivable. They then use these assets to provide goods and services to their customer members that involves both fixed and variable costs. If the cooperative is successful in these processes and operates efficiently, it generates a profit. Generating a profit is not automatic and, in fact, can be very difficult. Although minimizing the challenges in creating a profit, we will assume that the cooperative has a profit, also termed *net savings*, and discuss the alternatives for using those funds.

After generating a profit, the next set of critical financial decisions becomes how to distribute that profit to the patrons. The cooperative also needs to retain a portion of the profit to fund reinvestment in plants and equipment and fund equity retirement. The process of profit distribution is therefore intertwined with profit retention. The cooperative board of directors has the authority and responsibility to decide how to distribute profits. The board often receives advice from the manager and chief financial officer, but the final decision rests with the board. The board would also need to consider any loan covenants, priorities specified in the bylaws, or restrictions specific to state or federal tax law.

Importance of Cash Flow

In discussing profit distribution and retention, it is essential to understand the concept of cash flow. *Cash flow* is the movement of cash in and out of the cooperative, and the cash flow balance summarizes the amount of cash available. Many students think of cash flow and profit as analogous concepts, but they are distinct. *Profit* represents the financial gain a company made during a period of time. Profit is basically the difference between the amount earned and the amount spent in producing something. A firm's cash flow is influenced by the facts that not all revenue and expenses are received in cash and not all cash expenditures and receipts are classified as expenses or income. As we have mentioned, depreciation is a major expense that does not require cash. On the other hand, loan principle payments, asset purchases, equity retirement payments, and cash patronage require cash but are not classified as expenses. The cooperative board must simultaneously balance the amount of profit they return to the members, the amount they reinvest in the cooperative, and the amount of cash they give to the members versus the amount channeled for other uses by the cooperative.

The cooperative's options for profit distribution have a direct impact on cash flow. Paying cash patronage obviously requires cash and therefore decreases cash flow. Since the cooperative is distributing stock and not cash, equity patronage preserves the cooperative's cash flow. The members receive equity (stock certificates), and the cooperative retain the cash. We will see in a future chapter that profit distribution affects taxation, which creates another cash flow impact. In general, when the cooperative passes taxation to the member level, it creates a deduction at the cooperative level and thus reduces its tax payments while preserving cash flow.

Decisions in Distributing Returns

A number of intertwined decisions are involved in distributing net income and can be approached in a different order. For the sake of simplicity, we will order the choices and work through the decision process. The choices are summarized in Figure 12.1 at the end of this chapter.

Separating Member and Nonmember Profits

Member profits in a cooperative are taxed under Subchapter T, which is the cooperative specific section of the Internal Revenue Service (IRS) tax code. Nonmember profits are typically taxed differently and basically taxed at the standard corporate tax rate. We will see in a later chapter that there is one special form of cooperative (Section 521) that can treat member and nonmember profits in a similar fashion. The first step in distributing returns would be to separate member and nonmember profits. Because we are discussing the board of directors' decisions in distributing returns, most of which relate to member-based profits, we typically skip the initial step of separating member-based profits and discuss only the decisions related to distributing those profits. Nonmember-based profits will generally be retained as unallocated retained earnings, so we will skip discussing those profits for now.

DECISION 1: WHICH PORTIONS TO ALLOCATE TO MEMBERS AND TO RETAIN AS UNALLOCATED RESERVES

The first step is to decide what portion of (member-based) profits should be retained as unallocated retained earnings, also called *unallocated reserves*. The remainder will be allocated (placed in a member's name). The allocated form can be in a number of categories, including cash, qualified stock, nonqualified stock, and dividends paid on capital stock. The distinction is that all these categories represent specific amounts held by or given to specific members. Unallocated retained earnings functions as a general reserve account. The cooperative members have a collective claim on unallocated retained earnings; however, as the name implies, it is not apportioned to particular members and is never revolved or redeemed to the member. While we can discuss retaining funds as unallocated retained earnings, it is actually the default outcome. Any profits not distributed to the members in patronage or stock dividends end up as increasing unallocated retained earnings. More precisely, the cooperative has to pay taxes on these profits, so the default outcome is to add the after-tax portion to the unallocated retained earnings.

Cooperatives need an adequate amount of unallocated retained earnings as a *cushion fund*. Because cooperative equity is held on the balance sheet at face value, there is the expectation that the cooperative has the ability to repay it. If a cooperative experiences a loss, it does not have the equity to repay all its debt and also redeem all outstanding equity. So, if a cooperative experiences a loss, some category of equity must be reduced. In the absence of unallocated retained earnings, the members' stock would have to be written down in value in any year with a loss. Unallocated reserves provide the cushion that can be reduced in a loss year without writing down stock. Profits retained as unallocated are never returned to the member, so channeling profits to unallocated retained earnings reduces the members' return from the cooperative.

DECISION 2: WHICH PORTIONS OF THE ALLOCATED AMOUNT TO PAY AS DIVIDENDS ON CAPITAL STOCK AND AS PATRONAGE (CASH AND STOCK)

Patronage refunds are the distribution of profits in proportion to use or business volume. Since we are limiting this discussion to member-based profits, we can ignore nonmember issues. This decision therefore comes down to the question of whether to distribute profits to members in the form of patronage or dividends on stock. If the answer is patronage, there are more subcategories to decide upon.

Dividends on capital stock reward ownership rather than use. Historically, dividends has been a very minor aspect of profit distribution for U.S. cooperatives. As you might recall, Rochdale principles stated that dividend payments should be limited to no more than 8%. That level has made its way into several laws that affect cooperatives. In practice, most cooperatives do not pay any dividends on stock. Perhaps the U.S. cooperative industry has overreacted to the Rochdale pioneers caution against overemphasizing the return on ownership. It is true that distributing and profits as dividends reduces the potential amount of patronage refunds. In any case, we should recognize paying dividends on capital stock as one of a cooperative's options for profit distribution.

The majority of allocated income in a cooperative is distributed as some form of patronage. In determining patronage refunds, the board must determine the unit (physical or monetary) and number of categories. The best guideline is to use the unit and number of categories that best reflect the usage. Physical units are preferred when costs and benefits are related to volume. For example, most grain marketing cooperatives calculate patronage on a per-bushel basis. The cooperative's cost are related to volume, not price, and most grain handlers price grain to maintain a constant per-bushel profit margin. Monetary units are preferable when the costs and risks relate to the dollar amount (e.g., a credit cooperative) or physical units are not easily distinguishable. For example, a farm supply store carries hundreds of items, so it would be difficult to define volume in terms of physical units. Patronage can be calculated separately for products or departments or combined. In determining the number of patronage pools, there is a tradeoff between fairness and complexity. A single pool can result in one product or service being subsidized by other areas. On the other hand, multiple pools require more recordkeeping and can be confusing to the members. There are some overhead costs, such as the salary of the general manager and the office expenses that are related to all business areas. These overhead costs must be allocated or apportioned across the various department and activities. The calculations of profits and patronage in subcategories can therefore never be perfect but rather are affected by how overhead costs are allocated. The major rationale for separate patronage pools is when some products or departments are used only by a portion of the membership. If all the patrons use all departments and products, the issue of patronage pools is of less importance.

DECISION 3: DIVIDING PATRONAGE REFUNDS INTO CASH AND STOCK

The third decision is regarding what percentage of the patronage refund portion to pay in stock and what portion to pay as retained patronage (stock patronage). Cash patronage provides an immediate benefit to the members. In terms of the choices of return distribution, cash patronage increases the members' return more than any other alternative. Cash patronage is taxable to members, so the members are actually gaining the after-tax portion. However, almost all activities that generate cash for the producer are taxable, so members do not object to paying taxes on cash patronage. Cash patronage is still much more beneficial relative to some forms of retained patronage (stock patronage) that are also taxable. It is easy to see why a member would prefer receiving cash even if they had to use some of it to pay the tax obligation, relative to receiving equity, in which they have to use cash from other sources to pay the tax and then wait years for it to be revolved into cash. Members would obviously prefer a cooperative to distribute 100% of profits as cash patronage. The problem from the cooperative's standpoint is that the cooperative needs to retain some profits to fund infrastructure reinvestment and equity revolving payments. If we were ranking choices in terms of the cooperative's cash flow, cash patronage is the least desirable. If a cooperative issues a portion of the refund in the form of qualified stock, IRS regulations require it to pay at least 20% cash patronage. The logic is that the members should receive enough cash to pay the taxes on the qualified stock patronage. In practice, many members face tax rates higher than 20%, and most boards feel that they should pay 30%–40% cash patronage if they are issuing qualified stock patronage.

Cooperatives can also make patronage refunds in the form of retained patronage, also called *stock patronage*. Retained patronage is allocated because it is issued to individual members in specific amounts. Therefore, a member's allocated patronage can be in a combination of cash and retained patronage. Years ago, cooperatives issued patronage in the form of stock certificates, creating the term *stock patronage*. Today, most cooperatives simply track equity patronage in their accounting system. The equity is described as retained patronage because the cooperative is retaining the profits by distributing equity instead of cash. Describing a cooperative as distributing retained patronage sounds like a contradiction. For that reason, it is less confusing to describe retained patronage as stock patronage. Cooperative boards need to consider both the cooperative's cash needs and their need for equity on the balance sheet in determining the portion of returns to issue in the form of cash patronage.

DECISION 4: ISSUING STOCK PATRONAGE AS QUALIFIED OR NONQUALIFIED

Stock patronage can be in the form of qualified or nonqualified stock. The term *qualified* refers to whether the retained patronage amount qualifies as a deduction from the cooperative's taxable income in the year issued. Qualified stock is deductible for the cooperative when issued, and it is also taxable income for the members when issued. Cooperatives can issue stock patronage in another form, called *nonqualified*. Nonqualified stock patronage does not create a tax deductible for the cooperative or taxable income to the member in the year issued. Instead, it is tax-deductible to the cooperative and taxable to the patron during the year when it is redeemed into cash. In either case, the cooperative eventually receives the tax deduction, and the member eventually pays the tax. When the cooperative distributes qualified stock, it receives the tax deduction immediately in the year when the stock patronage is issued. Retaining profits by issuing qualified stock is therefore the most beneficial option for preserving the cooperative's cash flow. The cooperative retains all the profits associated with the qualified stock portion since it has no tax liability on them. From the members' perspective, it ranks near the bottom in terms of improving member return since the member immediately has to pay the taxes on profits they receive in the form of stock.

There are a number of IRS requirements for a stock patronage distribution to be qualified. The cooperative has to pay at least 20% cash, the patron has to receive written notice of the allocation, and the patron has to have agreed to accept the tax liability (which is typically part of the membership application). If the stock patronage does not meet those requirements or the cooperative does not choose to treat the stock patronage as qualified, it is automatically treated as nonqualified.

Ranking the Alternatives

The cooperative board of directors faces a balancing act in protecting the cash flow and financial stability of the cooperative on the one hand and providing benefit to the members on the other hand. In the long run, we would hope that there is not a conflict. Protecting the cooperative ensures that it will be there to provide patronage in the future. Profit distribution highlights the inherent balancing act. In terms of member benefits, the alternative ranks from most to least desirable cash, nonqualified stock, qualified stock, and unallocated equity. Members benefit from getting cash and getting it sooner and deferring taxes when possible. In terms of the cooperative's cash flow, the cooperative benefits from unallocated equity, qualified stock, nonqualified stock, and cash (the exact opposite order). The cooperative would prefer unallocated equity because it keeps the cash and never returns it. Next, it would prefer qualified stock because it retains the profits and passes the taxation on to the patron. Next, the cooperative would prefer nonqualified stock. Because nonqualified is not deductible when issued, the cooperative has to pay the tax and retains the after-tax portion. It will get the tax deduction in a later year when it is redeemed. Finally, issuing cash patronage (the most desirable for the members' return) is least desirable for the cooperative's cash flow.

How Should the Board Make the Decision?

The cooperative board needs to consider both the cooperative balance sheet and its cash flow needs as well as the members' need for a return. Unallocated retained earnings serves as a cushion fund that can be used to absorb a loss without writing down the value of the revolving equity. A common recommendation is for boards to build enough unallocated equity to absorb one or two years of foreseeable loss. The board should also consider what percent of the total equity it believes should be allocated to members to maintain a sense of ownership by them. Historically, many cooperatives indicated that they never wanted to exceed 50% of their equity being unallocated. In recent years, the percentage of unallocated equity has increased. Hopefully, that has been a well-thought-out decision and not a byproduct of cooperatives having available tax credits, which makes it easier to retain funds as unallocated retained earnings.

The decisions of the cash and stock portion and whether the stock portion should be qualified or nonqualified should be evaluated based on the cooperative's need for cash flow and additional equity. When the board redeems previously issued equity, that reduces both the cooperative's cash and equity. This can create a need to distribute new profits as stock patronage. The board always consider the cooperative's cash flow and equity needs but should be focused on maximizing the members' long-term return from the cooperative as its overall goal.

Allocating Losses

Despite the best intentions and planning, cooperatives, like other firms, can experience a loss. The alternatives for handling losses are somewhat different in cooperatives versus investor-owned firms (IOFs). IRS regulations and cooperative principles influence how losses are handled. Cooperatives can net losses across departments and products. That creates a tradeoff for the practice of multiple patronage pools. Multiple pools are generally fairer but increase the chance that a loss will occur in some area.

Reduce the Value of Unallocated Retained Earnings

A common alternative for allocating a loss is to reduce the value of unallocated equity. This alternative does not require any specific action of the members or decrease the value of their stock. In fact, having the ability to absorb a loss without a stock write-down is the primary rationale for holding unallocated retained earnings.

Write Down the Value of Member Stock

Another alternative for allocating a loss is to reduce the value of the member stock. This requires a decision of which members to allocate the loss to. In an IOF, losses are allocated in proportion to ownership. IOFs do not have to reduce the value of stock when a loss occurs since the stock trades on the market. The effect on stock value comes through the market price. The stock price decline affects owners in proportion to their ownership. In a cooperative, losses are allocated in proportion to past business volume using some sort of look-back period. Typically, the cooperative defines a *look-back period* as six to 10 years and allocates the loss in proportion to the business the members did during that look-back period. The goal is to distribute the loss to the members who used the products and services associated with the loss. The cooperative structure tends to allocate the loss to active members rather than to inactive members who have only an ownership stake.

Since cooperatives have historically issued qualified stock, the members have already paid tax on the profits, which is reflected by their revolving equity. Writing down the value of qualified stock gives the member a taxable loss that they can use to offset other taxable income. A stock write-down is therefore more beneficial from a tax standpoint hen reducing unallocated retained earnings. The disadvantage of a stock write-down is the publicity and member relations. Qualified stock has never been popular with members since they are paying tax on stock that is not yet cash. Some cooperatives have long revolving periods, which adds further to the negative impression of the cooperative's equity. If a cooperative is forced to write down their stock value, members are quick to complain that the revolving stock is risky and may never be redeemed. It is easy to see why a board would like to avoid the publicity of writing down stock value.

A stock write-down also creates recordkeeping challenges. Members have a portfolio of stock shares earned during various years. When stock earned in one or more years is written down, these stock shares must be adjusted to partial value. When additional stock patronage is issue, these shares are not readjusted to full face value; rather, additional shares are added. If a member held stock for 20 years until it began to revolve and then several write-downs occurred, it is easy to imagine the member's confusion about the actual amount of stock they own.

As an aside, cooperative bylaws often specify that a member must hold one share of membership stock to have a voting write. For that reason, the value of membership shares (which represent a small fraction of total investment in an open-membership cooperative) is generally not reduced in a stock write-down. Preferred stock is also generated from direct investment and not created out of the profit stream. The value of preferred stock would also not be subject to a stock write-down. Of course, dividends on preferred stock are subject to the board's discretion and would generally not be paid when the cooperative experiences a loss. The board can suspend dividends on preferred stock, which is what distinguishes it from debt. If the board issued the preferred stock with a stated dividend, it could suspend the dividend when required by financial conditions, but it could not pay patronage if it did not pay the dividend. That preference for the order of profit distribution relates to the term *preferred.*

Summary

One of the most important decisions made by the cooperative board of directors is how to distribute the firm's net income. The first step is to separate member and nonmember profits. In focusing on the member profits, the board must make four decisions.

First, they must decide which portion to allocate to the patrons and which to retain as unallocated retained earnings. Second, they must decide which portion of the allocated amount to issue in dividends on capital stock and which to issue as patronage. Third, they must decide which portion of the patronage to issue in cash and which to issue as stock patronage. Finally, the board must decide whether to issue stock patronage as qualified or nonqualified stock. In making these decisions, the board has to balance the members' need for a financial return with the equity and cash flow needs of the cooperative. If a cooperative experiences a loss, the board must make a similar decision as whether to allocate the loss to the members or reduce unallocated retained earnings. If the loss is allocated to the members, the cooperative establishes a look-back period and reduces the value of stock earned during that period.

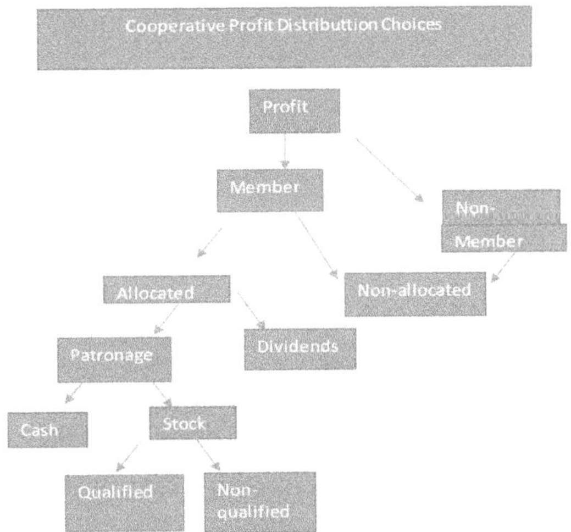

CHAPTER 13

Managing Capital Structure

Introduction and Learning Objectives

Cooperatives finance all their assets by some combination of debt and equity. Those assets include the fixed assets, such as the plant, building, equipment, and rolling stock, as well as the short-term assets, such as accounts receivable and inventories. When we discuss capital structure, we are referring to the choices that the cooperative made in creating the capital to fund all its assets. Capital structure affects the firm's profitability and return on investment (ROI) and also has major impacts on risk. This chapter will give you a better understanding of a cooperative's choices in funding all its necessary assets.

Financial Leverage

There are two primary sources in funding a business: debt capital and equity. The degree to which the cooperative is funded by debt is measured by either the debt-to-equity ratio or the debt-to-asset ratio. Both ratios are measures of financial leverage. A firm that funds a greater portion of its assets with debt is said to be more highly levered. Financial leverage amplifies the effect of profit or loss on the owners' return. When the firm is highly leveraged (i.e., has a higher portion of debt financing), a profitable situation will result in greater impact on the owners' ROI relative to a similar firm that experienced the same profits but was funded only by equity. On the other hand, when the firm is highly leveraged, an unprofitable situation will result in greater losses to the owners than that of an unlevered firm. Debt capital increases the owners' return only when the internal rate of return of the cooperative is greater than the interest rate. Cooperatives with higher profits are therefore more likely to have greater leverage.

The Nature of Debt Financing

Debt is money borrowed from a lender with a promise to pay the principal and interest on a set schedule. Almost all loans involve a loan contract, which gives the lender certain rights. Failure to make the scheduled payments can result in the lender exercising those rights. The most extreme case would be for the lender to force the firm into bankruptcy and liquidate the assets. In that case, the value of the firm would be distributed to the lenders holding claims on the firm in the order of priority of those claims. Any residual value would then go to the firm's owners. It is that structure of debt financing that creates both the returns and risk of leverage. Because debt payments are fixed, if the firm is able to use the funds on projects that create profits in excess of the loan payments, then that excess flows to the owners and increases their return. On the other hand, if the firm is not profitable, the firm must still pay the lenders at the agreed-upon schedule, and the burden of those payments falls on the owners.

There is generally a rationale for a cooperative—or any other firm—to take on debt financing. First, as described, debt financing creates the opportunity for the owners to increase their ROI. Second, the owners of the firm often have difficulty in providing the funds to finance all the firm's assets. This is particularly true in agricultural cooperatives. The cooperative members are financing their own farming operations. The equity that they provide to the cooperative therefore competes for the funds they need to finance their farming operations. Cooperatives are also unique in that they are usually limited to obtaining equity capital from their user-members.

Open-membership cooperatives typically create that equity out of the profit stream by distributing a portion of profits in the form of stock. That means that they can only create equity when they are profitable. Closed membership cooperatives create equity by selling membership stock during their formation period. That membership equity is typically linked to a usage right. Equity drives in a closed membership cooperative are complex and costly. Those sources and structures for raising equity in cooperative firms are in contrast with investor-owned firms (IOFs) that can sell equity at any time. Most large IOFs are publicly traded, meaning that they can raise equity capital by selling stock on public stock exchanges. It is easy to see that cooperatives do not have unlimited access to equity capital and have a rationale for some degree of debt financing.

A third rationale for debt financing is that interest payments are tax-deductible. Cooperatives can reduce their taxable income by issuing cash and qualified stock patronage to members and redeeming previously issued nonqualified stock. Because of those tools to reduce taxable income, deductible interest expenses are not as important for cooperative firms. Still, most cooperatives have nonmember business, which creates taxable income. When a cooperative has taxable income equal or greater to its interest expense, the interest expense is reduced by the tax savings. For example, a cooperative paying 5% interest with a 30% tax rate would have an after-tax interest rate of 3.5%.

Cooperatives' Use of Debt

According to the most recent U.S. Department of Agriculture agricultural cooperative statistics (2019) (USDA, 2021), agricultural cooperatives financed 54% of their total assets with some form of debt. Larger cooperatives tend to have higher portions of debt financing. The smallest category of cooperatives (less than $5 in annual sales) had a debt-to-asset ratio of 36%. For all size categories of cooperatives, short-term financing represented more than 50% of total debt. Short-term financing is typically used to fund inventories, accounts receivables, and other short-term assets.

Sources of Debt

Commercial Banks

Cooperative banks (discussed in the following section) are an important source of both short- and long-term debt financing for cooperatives. Like other firms, cooperatives can also borrow from commercial banks. Historically, banks have not been active in financing cooperatives because they do not understand the business model. Commercial banks have a particularly difficult time in understanding cooperatives' structure of patronage refunds and equity redemption. The dynamic nature of cooperative equity is troubling to bankers who are used to dealing with permanent equity capital.

Cooperative Banks

The greatest real advantage cooperatives have in the acquisition of debt capital is that they can borrow from their own cooperative lenders. Agricultural cooperatives have access to CoBank in the Farm Credit System (FCS). Nonagricultural cooperatives can use the National Cooperative Bank (NCB), and rural utility cooperatives can use the National Rural Utilities Cooperative Finance Corporation (CFC). Because these lenders are themselves cooperatives, they understand cooperative finance better than other lenders. CoBank is part of the FCS (which is itself a cooperative) and has a national charter to serve cooperatives. By its original authorizing legislation, the FCS had 13 banks for cooperatives that were established regionally around the United States. In the late 1980s, CoBank was formed through the merger of 11 of the original 13 banks. By mid-1999, all the original banks had joined CoBank, which now has the sole FCS charter to serve cooperatives. The funds that CoBank lends cooperatives are borrowed from the government-sponsored enterprise market through the Federal Farm Credit Banks Funding Corporation (FFCBFC). In essence, CoBank sells bonds in national capital markets and then lends those funds to its cooperative borrowers. Although FFCBFC bonds are not backed by the U.S. federal government, the agency is associated with the government. For that reason, investors perceive the bonds to be low-risk and will purchase them at slightly lower interest rates relative to those of corporate bonds. This source of funding provides a relatively low-cost source of financing for cooperatives. CoBank is owned by the cooperatives that borrow from it. When a cooperative obtains its first loan from CoBank, it must purchase stock just like any member joining a cooperative.

Ultimately, the borrowing cooperative must make an investment in CoBank in proportion to the size of its loan. The borrowing cooperative most often acquires its equity via retained patronage refunds. The relationship between CoBank and its member cooperatives is equivalent to that between other cooperatives and their members. For this reason, it also has some of the same challenges in its relationships with borrowers that cooperatives have with their member patrons. Based on the legislation authorizing the FCS, CoBank has strict policies that regulate borrower eligibility for its loans. In order to eligible to borrow from CoBank, a business must:

1. Conduct 50% of its business with members
2. Have 80% of its equity owned by producers
3. Operate on a cooperative basis with respect to voting and profit distribution

Cooperatives borrowing from CoBank must also meet the financial requirements established by the bank. These requirements vary according to prospects for individual cooperatives. CoBank evaluates applications in much the same way as do other lending agencies. Such factors as repayment ability, collateral, balance sheet and income statement changes or trends, quality of management, and member support enter into approval of a loan. CoBank makes seasonal and long-term loans. The former are used to finance short-term seasonal needs such as inventories, which, in agribusiness, may represent a relatively large portion of total liabilities at certain times during the year. Long-term loans are used to finance long-lived assets, such as land, buildings, and equipment.

The National Cooperative Bank

The NCB was established by Congress in 1978 and then privatized in 1981. The NCB serves organizations that are defined as cooperatives as well as those having similar organizing principles. The scope of its membership is large and includes independent grocery cooperatives, employee stock ownership plans, housing cooperatives, purchasing cooperatives, and other similar groups. The NCB goes directly to the capital markets for debt financing and sells pools of cooperative loans into the market.

The National Rural Utilities Cooperative Finance Corporation

The CFC is a cooperative banking entity created by rural electric cooperatives (RECs) in 1969 to supplement government lending programs. Today, the CFC offers a full range of financial products and services to member utility cooperatives. Like the NCB, the CFC is a conduit for its members to the private capital markets. In addition to serving RECs, the CFC operates several affiliated organizations, including the Rural Telephone Finance Cooperative and the National Cooperative Services Corporation. All the cooperative banks, CoBank, the NCB, and the CFC, have one thing in common: They have special insights into the financial needs and challenges of the cooperatives that are their members. Member cooperatives know that they can count on these institutions when other private sources of borrowed capital may not come through with required funds or may not understand the needs of cooperative organizations.

Loans From Members

Cooperatives can potentially borrow funds from their members. Member financing raises numerous issues. One question is whether the cooperative will be considered to be selling a security. If so, the firm may have to register the offering with the U.S. Securities and Exchange Commission (SEC). That registration process is complex and may cost hundreds of thousands of dollars. Member financing also exposes members to an additional source of risk from the cooperative. The attractive aspect of member financing is that the interest rate that the cooperative pays is often higher than what members can obtain through their bank savings accounts. That creates the opportunity for the cooperative to obtain funds at a lower interest rate while giving members higher rates than those of their local bank. Historically, many local cooperatives made the mistake of issuing demand notes to patrons. The notes functioned like certificates of deposit in a local bank except that the members could cash them in at any time. When the economy took a downturn, the members all rushed in to redeem their notes at the exact time that the local cooperatives were struggling to obtain funds.

Trade Credit

Just like other businesses, cooperatives often finance a portion of their current assets with *trade credit*. When a cooperative purchases fuel, fertilizer, and other inventories, it often has a short period to make the payments. Those obligations show up as accounts payable on the cooperative balance sheet and are a source of financing. Regional cooperatives may be particularly willing to supply trade credit to local cooperatives.

Cooperative Equity

Equity is the investment that member-patrons make in the cooperative. According to cooperative principles, members have a key responsibility to provide equity. Cooperative equity can be allocated (issued in the specific name of a member) or unallocated (a general reserve fund that is collectively owned). Cooperative equity can also be revolving, semipermanent, or permanent. Revolving equity can be qualified or nonqualified. The distinction between qualified and nonqualified equity relates to the taxation of the patronage distributed as equity. Both qualified and nonqualified equity are forms of revolving equity and similar in terms of the cooperative's capital structure or equity management system.

Because cooperatives distribute profits based on use and not in proportion to equity ownership, there is little incentive for members to invest equity capital. Open-membership cooperatives often create equity out of the profit stream by issuing a portion of patronage in the form of equity. Closed-membership cooperatives often link each unit of equity with a usage right. Members hold equity in a closed-membership cooperative to use the cooperative and receive patronage in proportion to their use.

Equity in any firm is risk capital. Equity holders have the right to the residual returns. In other words, the owners of any firm have a claim on what is left over after all the creditors have been paid. When a cooperative is profitable and retains a portion of the profits as equity, then the equity value of the cooperative increases. If the cooperative loses money, then the equity value of the firm decreases. Lenders like to see a strong equity base because it enhances their collateral position and makes the loan less risky. When a substantial portion of the assets are financed by equity, the lender knows that if the cooperative is unable to make the loan commitments and forced into bankruptcy, it is likely that the assets can be sold to generate enough funds to pay off the creditors. The cooperative equity is the basis for its borrowing capacity. Agricultural cooperatives have historically financed between 40% and 50% of their assets with equity.

Direct Investment

Direct investment is equity created when a member makes a cash contribution or investment in exchange for equity. Direct investment is allocated equity since it is held in the names of specific patrons. This type of investment is semipermanent. It is typically not redeemed by the cooperative until the death of the member. The most common form of direct investment is the membership share of stock that a member must purchase as a condition of joining the cooperative. In an open-membership cooperative, the membership share is often a nominal amount ranging from $100 to $1,000. In closed-membership cooperatives, direct investment shares of equity have associated usage rights, and members must purchase shares in proportion to their use of the cooperative. For example, when Dakota Growers Pasta Company was formed, members bought shares of stock in the cooperative for $3.85 per share. Each share purchased required the member to deliver one bushel of durum wheat each year. The cooperative raised $12.5 million in equity in this manner and secured for itself nearly 3.25 million bushels of durum. For new-generation cooperatives, the vast majority of their equity is in the form of direct investment.

Direct investment is usually the only option available to raise the equity to start a cooperative. The equity drive to obtain the direct investment helps determine whether there is enough interest and commitment to establish the cooperative. By raising direct investment, the incorporating board can ensure that the cooperative has the proper level of equity. In terms of existing cooperatives, direct investment or membership stock ensures that members have some skin in the game before they are granted the right to vote and use the cooperatives' assets. As the amount of direct investment is increased, the cooperative has less need to retain profits as equity and can issue higher portion of cash patronage. The major disadvantage of direct investment is that because cooperative profits are distributed in proportion to use, there is no return linked to investment. In an open-membership cooperative, equity does not appreciate and cannot be bought and sold. Because of that, members have little interest in investing in cooperative equity. In a closed-membership cooperative, there may be a limited market for the cooperative equity because of the usage rights. Direct investment is not generally a recurring source of equity capital since it takes a special effort to develop an equity drive, which often requires careful compliance with security laws and regulations.

Retained Patronage Refunds

Retained patronage refunds are portions of the net income allocated to members but paid in the form of equity rather than cash. Retained patronage refunds create allocated equity since they are held in specific patron accounts. Retained patronage refunds are typically revolving equity, meaning that they are eventually redeemed for cash by the cooperative at their original face value. Traditional, open-membership cooperatives have traditionally raised the majority of their equity through retained patronage refunds.

The advantages of retained patronage from the members' perspective are that it was created from the profit stream and members do not have to make a cash investment. Members essentially earn their way into ownership. Retained patronage is well suited for supply cooperatives and grain marketing cooperatives. In those instances, producers have other options for purchasing inputs and selling their commodities. While they might be interested in joining the cooperative in order to be eligible for patronage, they are unlikely to make a large up-front direct investment. Retained patronage is a systematic method for the cooperative to build equity, one that it is relatively painless for the members.

The disadvantage of retained patronage from the cooperative's standpoint is that it is dependent on the cooperative's profitability. If the cooperative has a loss year, its equity value is reduced. That creates a need for more equity, but the cooperative obviously cannot created retained patronage equity until it is profitable. The revolving fund aspect of retained patronage also creates challenges for the cooperative board of directors. When the board redeems equity, the cooperative is using cash and also reducing (destroying) equity. This creates the need to create more equity and cash through retained patronage. Revolving equity creates a complex balancing act for the board of directors. Another disadvantage of retained patronage and revolving equity is that members may expect the cooperative to revolve equity regardless of its financial condition. Members realize value for the share of profits distributed in the form of equity only when that equity is redeemed for cash. That causes members to want the cooperative to revolve equity as rapidly as possible and to keep the revolving cycle constant or increasing.

Per-Unit Capital Retains

Per-unit capital retains, also called *per-unit retains*, are equity deducted from members' commodity payment for each unit of commodity handled. Per-unit retains are therefore based solely on the amount of commodity handled and not affected by the cooperative's profitability. Because of this, per-unit retains create a stable source of equity for the cooperative. They are also logical for fruit and vegetable cooperatives, in which commodity prices vary widely. Because profits vary widely on a day-to-day basis and across commodities, linking equity creation to profits may not be fair. The disadvantage of per-unit retains from the members' standpoint is that the cooperative deducts equity even when commodity prices are low. Some members may also feel that the stable source of equity, regardless of profitability, places less pressure on the cooperative to operate profitably and efficiently. Because per-unit retains are deducted from the commodity payment, some members perceive that the cooperative is not price-competitive. Per-unit retains create allocated equity and can be allocated as either qualified or nonqualified equity. They are taxed in a similar way to patronage refunds except that a cooperative issuing qualified per-unit retains does not have the requirement of issuing at least 20% in cash patronage, which is a requirement for qualified patronage refunds.

Unallocated Equity

As we have discussed, *unallocated equity*, also called *unallocated retained earnings*, is a general reserve fund that is collectively owned by the members but not individually owned. As the name implies, unallocated equity is unallocated and the only permanent category of equity in a cooperative since it is never revolved or returned to the members. One of the purposes of unallocated equity is to create a reserve fund that can be reduced if the cooperative experiences a loss. In the absence of unallocated equity, the cooperative would have to write down the value of the members' stock in any year in which a loss occurred.

A more controversial source of unallocated equity is when a portion of member profits is retained as unallocated equity. The disadvantage of retained member profits as unallocated equity (from the members' perspective) is that unallocated equity is never revolved by the cooperative and the members never receive those profits. If a cooperative is liquidated, the residual value of the firm, after all creditors are paid, is distributed to members in proportion to past use during some defined look-back period, typically about six years. Active members therefore still have a claim on the assets funded by unallocated equity. However, once a member becomes inactive, they eventually leave behind any claim on the unallocated equity. A high level of unallocated equity can therefore create an incentive for members to liquidate a cooperative in order to capture the unallocated equity's value. This phenomenon is known as *demutualization*. A high level of unallocated equity is rarely the direct cause for members to vote for liquidation but can be a contributing factor.

Preferred Stock

Most cooperative bylaws permit the cooperative to sell *preferred stock* to either members or nonmembers. Preferred stock owners do not have voting rights. If the cooperative is liquidated, the preferred stockholders have a claim on the residual value ahead of the common stockholders—hence, the origin of the term *preferred*. There is typically no market for preferred stock in a cooperative, and as we know, cooperative profits are distributed based on use, not on stock ownership. Because of these facts, cooperative preferred stock typically carries a dividend rate. The maximum dividend is often limited by law to 8%. The board has discretion not to pay the dividend if the cooperative does not have the financial resources. This is what makes preferred stock equity and not debt. However, the board cannot pay patronage unless the preferred dividends have been paid. Because there is no public market for cooperative preferred stock, cooperatives typically establish a system by which members can petition the board to have the preferred stock redeemed.

In recent years, some progressive cooperatives have redeemed revolving equity by converting it to preferred stock. The effect is that after some period of time, the cooperative begins to pay dividends on patron equity. That can be beneficial to the cooperative since it doesn't have to expend the cash to revolve the equity (it only has to make the dividend payment). Many members may also be happy to hold the equity since they are receiving a dividend. The cooperative can also use the profits from nonmember business to pay the dividend payment. The preferred stock dividend is often higher than the cooperative's interest rate. It is therefore not a cost-effective alternative for the cooperative, but since the funds are going to members and not to outside lenders, it may still be considered to be in the members' interest.

Nonmember Sources of Equity

One limitation of the cooperative business model is that it generally does not lend itself to raising capital from nonmember investors. Cooperatives can sell preferred stock to nonmembers. As mentioned previously, the dividends on cooperative equity are often limited to 8%, either by law or by the cooperative's bylaws. In low-interest-rate periods, cooperative preferred stock can be attractive to nonmembers. However, when interest rates rise, investors have alternative investment choices that are considered lower-risk and that provide a higher rate of return. Regional cooperatives are able to justify the security registration costs and issue preferred stock that is publicly traded on stock exchanges. Local cooperatives cannot justify that process, which means that there is no formal market for their preferred stock. That makes the preferred stock an illiquid investment, which also limits the interest from nonmember investors.

The hybrid member-investor cooperative form has two classes of members. Investor-members receive a share of profits based on their ownership, whereas user-members receive a share of profits based on patronage. That creates a direct incentive for nonmember investment and addresses some of the limitation in raising nonmember equity. However, unless the cooperative is large enough to afford to register with the SEC, there will probably not be a formal market for the investor-member stock. The investor-members may also be concerned that the user-members will manipulate the price paid for commodities or charged for inputs such that the users benefit while reducing the profits paid to the investors. For those reasons, the member-investor model has not been particularly successful as a vehicle for attracting nonmember investment.

Table 13.1 Classification of Cooperative Equity

	Allocated or Unallocated	**Revolving, Semipermanent, or Permanent**	**Qualified or Nonqualified**
Direct investment	Allocated	Semipermanent	Not applicable
Retained patronage	Allocated	Revolving	Can be either
Per-unit retains	Allocated	Revolving	Can be either
Preferred stock	Allocated	Semipermanent	Not applicable
Unallocated retained earnings	Unallocated	Permanent	Not applicable

Summary

Every asset that a cooperative needs to operate, from land and buildings to inventory and accounts receivable, must be funded by a combination of debt and owner equity. Debt financing is often referred to as *financial leverage*. Debt financing can increase owners' returns when conditions are favorable; however, it also increases the chance of negative return and bankruptcy when conditions are unfavorable. Cooperatives can obtain debt financing from commercial banks but often borrow from specialized cooperative banks that understand their business model.

Cooperatives are also financed with owner equity. Cooperative equity can be allocated (in a member's name) or unallocated and it can be semi-permanent or revolving. Closed membership cooperatives obtain most of their equity from direct investment in membership stock while open membership cooperatives create most of their equity out of the profit stream by issuing patronage in the form of equity. Retained equity patronage is not bought or sold but is instead eventually redeemed at face value by the cooperative. That creates the complex process of equity management as the cooperative is simultaneously creating new equity through retained patronage and also eliminating equity by redeeming equity for cash. That process of equity management is unique to the cooperative firm and is also the topic of the next chapter.

Reference

United States Department of Agriculture. (2021). *Agricultural cooperative statistics* (Rural Business Cooperative Services, Service Report No. 83). U.S. Department of Agriculture.

CHAPTER 14

Equity Management

Introduction and Learning Objectives

The structure and management of equity in a cooperative is unique from that of other business models. Most of the differences stem directly or indirectly from the fact that cooperatives distribute profits based on use and not in proportion to equity ownership. That creates various structures for equity, including the linkage of equity with usage rights in closed-membership cooperatives and distribution of patronage in the form of revolving equity in open-membership cooperatives. As the name implies, revolving equity is eventually redeemed into cash by the cooperative. Under this structure, the cooperative is constantly creating additional equity through equity patronage and eliminating equity by redeeming it in cash. That entire process is called equity management and is an important and challenging activity of the board of directors. Revolving equity is typically redeemed under a predetermined systematic process, and there are multiple alternative equity redemption systems. After reading this chapter, you will have a better understanding of the process of equity management and the advantages and disadvantages of alternative equity management systems.

Equity Redemption

Equity redemption is the process of returning cash to members who hold previously issued equity patronage. Equity redemption is unique to cooperatives. Because cooperatives distribute profits in proportion to use there is no direct rationale for holding equity. Because of that there is no market for cooperative equity. (The exception would be if the cooperative equity has an associated usage right.) Because cooperative equity typically has no market value, the cooperative must establish a system to eventually redeem the equity for cash. It is only through equity redemption that the member realized any benefit from the patronage distributed as equity. Due to the time value of money the member's return is increased when the cooperative revolves equity more rapidly. Shorter equity revolving periods reduce the cooperative's cash flow and thus create another tradeoff between the cooperative's financial stability and member benefit.

The process of equity redemption or equity management is also important to keep the members' investment proportional to their use of the cooperative. If a cooperative fails to revolve equity or has a long equity revolving period, many older members may have high equity balances but have slowed or discontinued use of the cooperative. These members are overinvested in the cooperative. On the other hand, younger members may use the cooperative but have relatively little equity and hence be underinvested. Ideally, a cooperative board would like to manage equity so that equity investment is roughly proportional to use. Failing to redeem equity reduces the member's return from the cooperative and also results in members perceiving little or no value in terms of stock patronage. State statutes generally require a cooperative to redeem equity if a patron dies. A systematic system of equity redemption therefore helps the cooperative avoid the unpredictable timing of estate settlements.

Equity Redemption Systems

For the abovementioned reasons, equity redemption is important to maintain the members' return from the cooperative, help keep equity in proportion to use, and eliminate unpredictable outflows to estates. On the other hand, equity redemption reduces the cooperative's cash flow as well as equity, which increases its debt-to-equity ratio. Equity redemption is therefore a complex balancing act for the cooperative board of directors. An equity redemption system is an established plan for redeeming equity. Most cooperatives have systematic systems that might be based on the age of stock, age of patron, percentage of all equities or base capital. The cooperative can also have a nonsystematic program such as one that allows members to petition for equity redemption in the case of hardship. Estate settlements are also often categorized as nonsystematic programs.

Factors Affecting the Choice of Equity Redemption Systems

Cooperative boards consider the following factors when selecting an equity redemption system:

1. They would like the system to be fair to members. In general, this means the system keeps equity investment proportional to the use of the capital. It is logical that members who make greater use of the cooperative should have a higher level of equity investment and that the cooperative should redeem the equities of inactive or low-volume members.
2. The equity management system should encourage members to contribute equity or at least support the structures that create equity. Members are more likely to support equity management systems that redeem equity more rapidly and consistently.
3. The cooperative must consider whether it has the cash flow to support the plan. A cooperative needs a high level of profits to be able to revolve equity consistently according to a rapid cycle. A cooperative's profit and cash flow also change year to year due to weather and other conditions. The cooperative board must consider whether the equity redemption system provides enough flexibility for the cooperative to adjust redemption payments according to available funds.
4. Local cooperatives would like the equity management system to be compatible with that of the federated cooperative. When the local cooperative receives patronage from the regional cooperative, that patronage is a combination of cash and regional cooperative stock. This becomes part of the local cooperative's profits that are passed on to members in a combination of cash and stock. In essence, the cooperative is issuing stock patronage to the members, some of which is really pass-through stock patronage from the regional cooperative. The local cooperative already faces cash flow issues if the cash and stock percentage does not match that of the regional cooperative. The same issue occurs when the local cooperative redeems equity. If the local cooperative redeems equity more rapidly than the regional cooperative, then it is issuing its members cash for regional equity, which it is still holding on the balance sheet. Similar problems occur if the local cooperative is on a different equity management system relative to the regional cooperative.
5. Finally, the cooperative would like the equity redemption system to meet the requirement of the lenders. The cooperative's creditors require it to maintain a minimum equity-to-asset ratio and also a minimum cash balance. Redeeming equity reduces both cash and equity. Equity redemption programs that are not tied to the cooperative's financial condition can therefore hinder a cooperative's ability to repay loans and limit future borrowing ability. Ideally, the cooperative board would develop an equity

redemption budget that is based on the cooperative's financial situation. They would then apply that budget to the specific equity management program. In reality, some types of equity management systems are more compatible with the concept of a redemption budget relative to others.

Comparison of Alternative Equity Redemption Systems

This section describes the alternative equity management systems and compares their proportionality, member benefit, redemption budget, and percentage of estate settlements. In determining proportionality, U.S. Department of Agriculture data on farming activity by age of operator were used to estimate the lifetime business volume of a typical cooperative member. That information was used to compare lifetime business volume with equity holdings. The member benefit value for the equity management plans was calculated as a ratio of the net present value of the redemption payments to the value of receiving the lifetime revolving patronage in cash. Receiving redemption payments over time obviously has a lower value relative to immediate cash payments, but the ratio reflects the delay in redemption. The cooperative's redemption budget was calculated as the percentage of annual profits needed to service the redemption plan. Finally, the percent of estate settlements was simply the percentage of a member's equity balance that remained unpaid at age 85. A summary table of the performance of the plans is presented after the plan discussion.

Age of Stock (Revolving Fund)

Under an age-of-stock plan, also called a *revolving fund*, the cooperative redeems the oldest equity on a first-in-first-out basis. For example, if the cooperative were on a 10-year age-of-stock plan, it would redeem the equity that it issued in 2009 in the year 2019. Cooperatives using an age-of-stock plan can attempt to maintain a fixed revolving schedule or can accelerate or slow the cycle according to the financial condition of the cooperatives. Cooperatives can also revolve part of a year or multiple years. For example, if the cooperative was very profitable in 2009 and issued a large amount of stock patronage, it might have to revolve that year over a two- or three-year period. At other times, it might have the funds to revolve two or three previous years in a single year. The age-of-stock system becomes more proportional as the revolving period decreases. Not surprisingly, member benefit also increases with shorter revolving periods. The cost of the improved proportionality and member benefit with a shorter revolving period is a higher redemption budget for the cooperative.

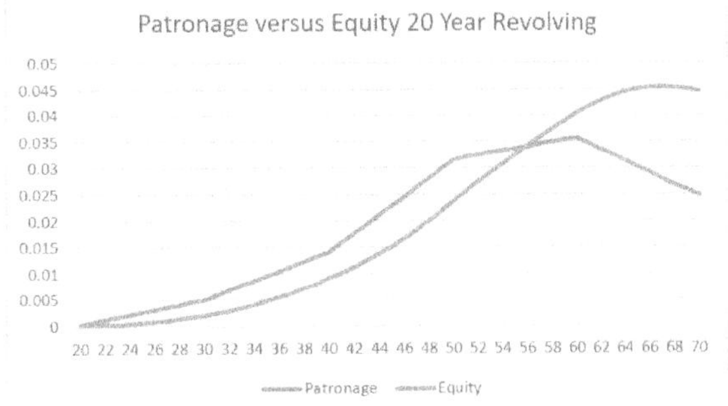

Figure 14.1 Patronage Versus Equity, 20-Year Revolving

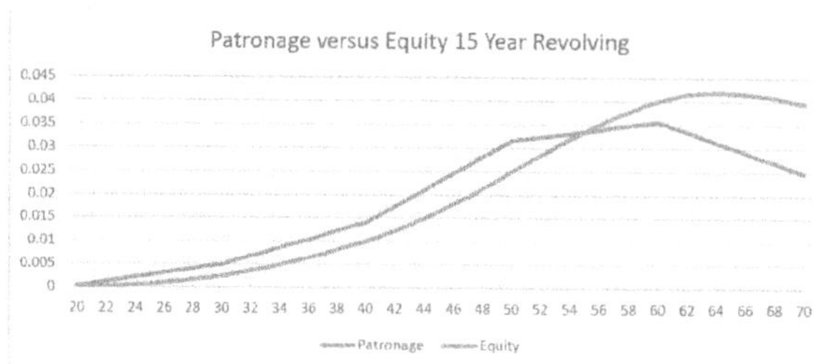

Figure 14.2 Patronage Versus Equity, 15-Year Revolving

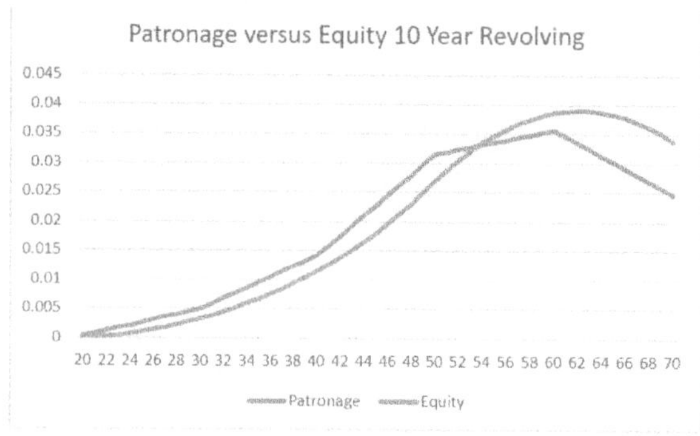

Figure 14.3 Patronage Versus Equity, 10-Year Revolving

Percentage of All Equities

Under the percentage-of-all-equities system, the cooperative redeems a percentage of every patron's equity balance each year. The board would establish the percentage based on the financial condition and balance sheet goals for the cooperative. For example, the cooperative may consider how much new equity was created through equity patronage and establish a percentage to redeem such that the total equity in the cooperative remains constant. The percentage of all equities programs is popular with new members because every member with an equity balance receives some redemption.

One disadvantage with the percentage-of-all-equities program is that it delays the time for a new, underinvested member to build equity in proportion to their use. Some consumer cooperatives use a percentage-of-all-equities program because they feel it helps new members understand how the equity system will work. When the new member receives a small redemption payment after one year, they may see how the redemption payments will increase as they build equity. The percentage-of-all-equities program never completely retires equity since the member receives only a portion of the balance each year. It therefore must be combined with some sort of lump-sum payment to close out the balance. As we can see from Figure 14.4, the percentage-of-all-equities program does not do a particularly good job of keeping equity in proportion to use. It rates relatively high in member benefit but also has a relatively high revolving budget and leaves a good portion of equity to be redeemed as estates. In general, a cooperative can achieve higher proportionality and lower estate settlements with an age-of-stock system than a percentage-of-all-equity system with the same budget.

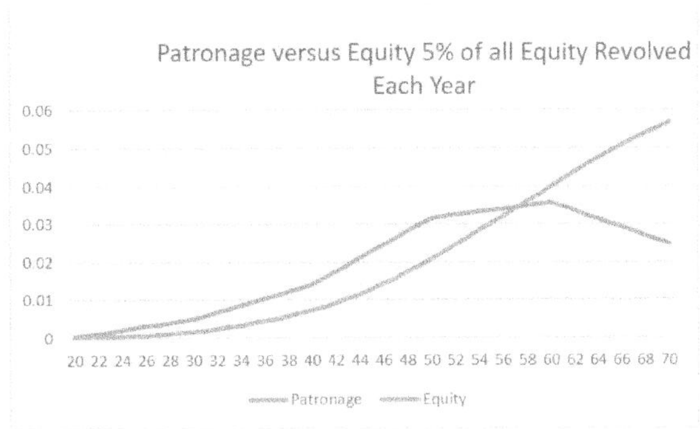

Figure 14.4 Patronage Versus Equity, 5% of All Equity Revolved Each Year

Age of Member

Under an age-of-member system, the member has all their revolving equity redeemed when they reach the specified trigger age. Many agricultural cooperatives were formed in the 1930s and 1940s. Their memberships tended to be young farmers who saw the need for the cooperative and were content to build equity to fund the cooperative's growth. The concept that the equity would be redeemed when they reached a retirement age was acceptable. Now that those cooperatives have matured, many of them struggle with the cash flow and equity implications of their age-of-member systems. A high percentage of their memberships are now nearing the redemption trigger age, creating challenges in meeting redemption cash flows.

Members also often come to expect redemption at a specified age to be a right or contractual obligation of the cooperative. That makes it very difficult for the board to adjust the trigger age to manage redemptions within a specified budget.

One of the characteristics of an age-of-member program is that the member becomes significantly overinvested prior to the trigger age and is then underinvested if they continue to use the cooperative. For that reason, reducing the trigger age does not necessarily improve proportionality. The equity left for estate settlement depends on the business volume that members conduct after the trigger age. As the trigger age decreases, the portion of estate settlements becomes larger. With today's age profile of farmers, an age-of-member program has a fairly high redemption budget while not ranking high in either member benefit or proportionality. Back when many cooperatives were started, the age-of-member system may have been attractive, but if we established a cooperative today, it is unlikely that this type of system would be adopted.

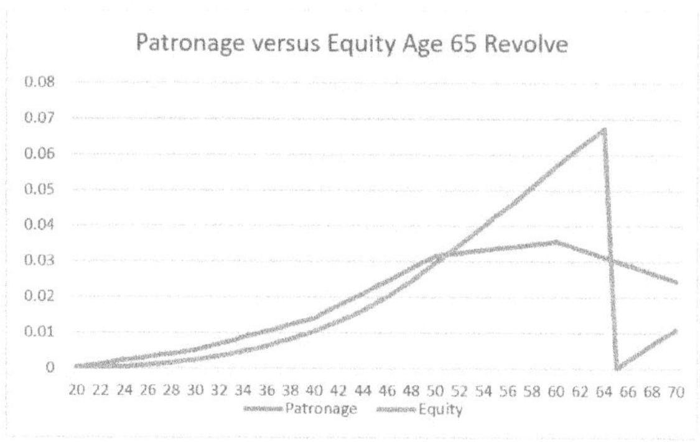

Figure 14.5 Patronage Versus Equity, Age 65 Revolve

Base Capital

The base capital system is conceptually very simple but can be complex to implement. Under the base capital system, the patron's percentage of the total equity in the cooperative is matched with the patron's percentage of the cooperative's business volume. In other words, a patron who did 5% of the cooperative business would be required to hold 5% of the cooperative's equity. Underinvested members have to invest more equity, while overinvested members receive equity redemption. The base capital system is therefore the best system for keeping equity investment in proportion to use. A base capital system is not attractive to new members since they would have to make an immediate investment to bring their equity balance up to the required level. A beginning farmer may not have the funds to invest the equity required under a base capital system.

In practice, many cooperatives using base capital do not require an underinvested member to make a direct investment but instead eliminate their cash patronage until they build the required equity level. Because those cooperatives do not instantaneously receive equity from underinvested members, they also cannot instantaneously redeem the equity of overinvested members. They therefore do not keep equity in exact proportion to use but keep equity balances adjusting toward that level.

In terms of our ranking, the base capital system is by far the most proportional. It is unusual that it has a very small redemption budget for the cooperative. Underinvested members are paying in, which funds the payouts to overinvested members. The base capital system has a fairly low member benefit rating since members are being forced to pay in earlier in their life. It would be expected to have a low percentage of estate settlements since most members would have become inactive and had their equity redeemed prior to their end of life.

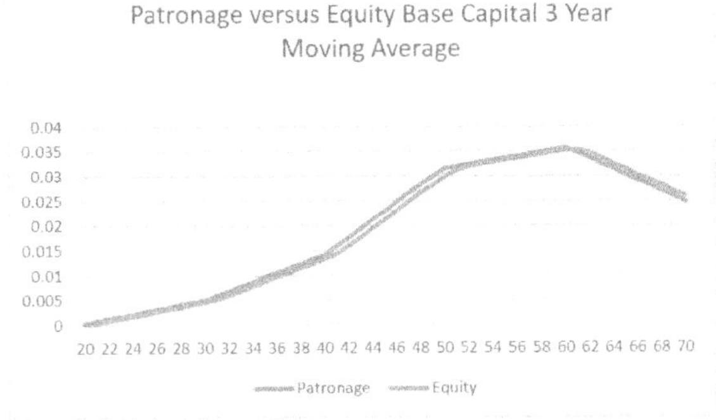

Figure 14.6 Patronage Versus Equity, Base Capital Three-Year Moving Average

Nonsystematic Methods

A *nonsystematic plan,* also called a *specialized plan,* is one in which equity is redeemed only if a special situation occurs. One special situation that has already been mentioned is the death of the member. Less frequent situations would be when the member moves away or is in a hardship situation. Aside from estate settlements that are often required by statute, cooperative board consider special situations on a case-by-case basis. Relying on a nonsystematic system would create a low financial burden for the cooperative. However, it would provide a poor financial return to members and result in unpredictable cash outflows. It also does a poor job of keeping equity in proportion to use. Nonsystematic or specialized systems generally do not meet any of the goals of a desirable equity management system. They tend to occur as the result of not implementing a systematic program.

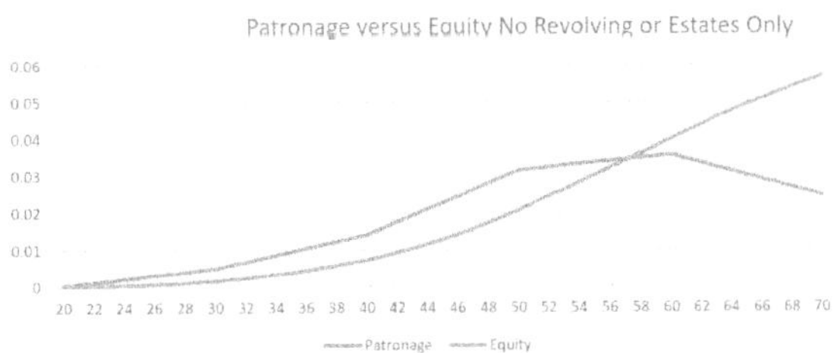

Figure 14.7 Patronage Versus Equity, No Revolving or Estates Only

Summary of Comparison of Equity Management Systems

Not surprisingly, no single system excelled in terms of all our desired properties. In general, there was a tradeoff between the system's performance in proportionality and member benefit versus the cooperative's redemption budget. This is most evident in age-of-stock programs that become more proportional and have a higher member benefit as the revolving period decreases. Those advantages came at the cost of a higher redemption budget.

The base capital system is by far the most proportional and has a low revolving budget; however, it also has a low member benefit ranking. Percentage-of all-equities systems have moderate member benefit but are not particularly proportional. The performance of age-of-member systems depends on the age profile of the membership. Although they might have had a low revolving budget (and also low member benefit) when cooperatives were started, cooperatives managing under these systems today find themselves with a high revolving budget and a system that does a poor job of keeping equity in proportion to use. A new young member would rank an age-of-member system as having a very low benefit. Of course, existing members in cooperatives who are nearing the trigger age likely support continuing these systems.

Alternatives to Redemption

While redeeming revolving equity is the method by which a member receives value for their stock patronage, it does create management challenges for the cooperative board. It is also difficult to design an equity management program that is attractive to members and meets the cooperative's needs for financial stability. Cooperatives have experimented with alternatives to equity redemption.

Redemption at a Discounted Rate

Under certain circumstances, some cooperatives have redeemed a member's equity before they would have otherwise received a payment but done so at a discounted rate. In essence, they paid the member the present, not the future, value of the equity. One situation for discounted redemption would be when a member of a cooperative with an age-of-stock program dies. In that case, the estate could wait and receive the equity over a number of years as each year of stock becomes eligible. In order to close the estate, the executor may ask whether the cooperative could pay the future redemption payments immediately but at a discounted rate to reflect the time value of money. If the equity balance is not large, the cooperative may be interested in complying in order to simplify its recordkeeping. Establishing a fair discount rate for discounted redemptions can be difficult, and some members do not understand the concept of the time value of money and therefore may consider the payment unfair. Most cooperatives describe the face value of the retained patronage as a benefit flowing to the member. They are therefore reluctant to discuss the present value of the revolving equity since it would be perceived as diminishing the value of stock patronage. A cooperative that implemented discounting would obviously want to specify the procedures and limitations carefully in the bylaws or policies since it would not have the funds to retire everyone's equity.

Conversion to Preferred Stock

As discussed previously, one alternative to redemption is to convert the equity to a dividend-paying preferred stock. That reduces the cooperative's immediate cash outflow, and the member may be content to hold the equity since they receive dividend payments. Unless the dividends can be paid with nonmember profits, the cooperative is reducing patronage-based profit distribution to issue a portion of profits to the preferred stockholders. The cooperative must also establish a system to eventually redeem the preferred stock or allow it to be sold to another member. Otherwise, they will eventually have to redeem the equity as an estate settlement.

Exchange of Equity Among Members

Closed-membership cooperatives often allow members to sell their equity and associated usage rights to another eligible producer. The price of the equity is negotiated between buyer and seller, although the cooperative board typically has to approve that the buyer is eligible for membership. Under this structure, the value of the equity and usage rights is determined by the anticipated amount of future patronage. It therefore tracks the cooperative's profitability. The possibility of exchange between members creates the potential for cooperative equity to appreciate in value. It also increases its liquidity since a member may have the opportunity to sell it at any time. Of course, by selling the equity, the member loses the right to use the cooperative. In addition, since the exchange is limited to other qualifying members, the market for the equity might be thin and result in an unfair price.

A few open-membership cooperatives have experimented with equity exchange between members. Similar to the concept of discounting, a member who desires immediate redemption could sell their equity to another member at a privately negotiated discounted rate. Again, another problem is that the market for the equity is thin. The member purchasing the equity also bears the risk that the cooperative expands the revolving period or writes down the value of the equity due to a loss.

Summary

A *cooperative capital structure* refers to its mix of debt and equity. Debt financing can increase the members' rate of return in times when the cooperative's internal rate of return (IRR) is greater than the interest rate but will decrease owners' return when the IRR falls below the interest rate. Most agricultural cooperatives finance a portion of their short-term and long-term assets with debt financing. CoBank and other cooperative banks are major providers of debt capital to agricultural cooperatives.

Cooperatives have unique methods of acquiring equity capital. Open-membership cooperatives tend to concentrate on retained patronage that becomes revolving equity. Closed-membership cooperatives concentrate on direct investment that is linked to a usage right and obtained during a formal equity drive. All cooperatives face constraints on obtaining equity capital since they can obtain capital only from user-patrons.

Cooperative equity can be allocated or unallocated. Direct investment, retained patronage, per-unit retains, and preferred stock are all allocated since they are held in the name of individual patrons. Retained earnings is unallocated and functions as a general reserve fund. Cooperative equity can also be revolving, semipermanent, or permanent. Retained patronage and per-unit retains are revolving equity; direct investment and preferred stock are semipermanent since they are typically redeemed at the death of the member. Unallocated retained earnings is the only category of permanent equity since it is never redeemed.

Cooperative must manage equity in order to keep it proportion to use and allow the patron to realize a return from the retained patronage. The most common equity management systems are age-of-stock, age-of-member, percentage-of-all-equities, and base capital. The age-of-stock system is the most common and can be fairly proportional at shorter revolving periods. In general, there is a tradeoff between the performance of the equity management system in terms of proportionality, member benefit, and avoiding estate settlements and the cooperative's revolving budget. The base capital system is by far the most proportion but is uncommon in local cooperatives due to the requirements for underinvested members to contribute equity.

CHAPTER 15

Cooperative Taxation

Introduction and Learning Objectives

Most students do not think much about federal and state income taxes. However, taxation is a major concern for businesses and for business owners. Cooperative taxation is unique in that it can shift taxation between the cooperative and the members. Because of that structure, taxation creates another balancing act between the cooperative's cash flow and the members' realized financial return from the cooperative. In recent years, cooperatives have taken on an additional role in pooling and distributing tax credits. This chapter will help you understand the principles and processes of cooperative taxation and the tax-related decisions of the cooperative board of directors.

General Principle of Cooperative Taxation

Firms that operate under the cooperative business model can qualify for special taxation under the Internal Revenue Service's (IRS) Internal Revenue Code (IRC). Cooperative taxation is therefore determined by the IRS and not by how the firm was incorporated. In general, the principle of cooperative taxation is that the cooperative is taxed as a corporation at the standard corporate rate but is able to deduct patronage distribution to members. Those patronage distributions are taxable to members, so the cooperative can shift taxation to them. The same holds true for both cash and stock patronage; however, in the case of stock (retained) patronage, these distributions can be structured so that the taxation is immediately transferred to the member when the stock is issued or later, when it is redeemed into cash. There is also a more restrictive category of cooperative, designated a Section 521 cooperative, that has special tax structures. Patronage from regional cooperatives also affects the taxation of local cooperatives.

Subchapter T

Any corporation or limited liability company (LLC) "operating on a cooperative basis" can be taxed under Subchapter T of the IRC. Subchapter T does not apply to mutual savings banks, mutual insurance companies, or rural electric or telephone cooperatives. All these organizations are taxed under separate special sections of the IRC. Consumer and retail cooperatives are taxed under Subchapter T. The taxation classification of a cooperative business is separate from its incorporation status. For that reason, an LLC or a corporation operating on a cooperative basis can be taxed under Subchapter T, whereas an entity incorporated as a cooperative would not qualify if it did not distribute patronage based on use and follow the other principles of operating on a cooperative basis.

The basic rationale of Subchapter T and cooperative taxation is that the cooperative is an extension of the patrons who own the cooperative. Subchapter T allows a cooperative corporation, in addition to making deductions for expenses allowed by other businesses, to also deduct certain distributions of net income made to the members. Those distributions become taxable income to the member. Consequently, the net income is taxed only once. In some cases, the taxation is immediately passed on to the patron, and in other cases, the income is taxed at the cooperative level, and then the cooperative receives a tax deduction and the member accepts the taxation for that income in a future period. Profits from nonmember business cannot be distributed as patronage under the general provisions of Subchapter T. Most cooperatives pay taxes on nonmember profits and any undistributed member profits at the general corporate rate. Subchapter T also includes Section 521 of the IRC, which describes the requirements for a more restrictive form of cooperative, termed a *Section 521 cooperative*. Section 521 cooperatives, which are discussed in greater detail later, must distribute profits to nonmembers but are also allowed to deduct that nonmember profit distribution.

Tax Effects of Patronage Refunds

Subchapter T allows cooperatives to exclude patronage refunds and per-unit retains from their taxable income. Patronage refunds are composed of profit distribution made to the patron on the basis of business volume. This volume can be calculated on the basis of quantity or value and separated by product or department or pooled. The IRS used the term *patronage dividends*, but most cooperative scholars prefer the term *patronage refund* to avoid confusion with a dividend made on the basis of stock ownership. With a few exceptions, nonmember profits and earnings not based on the use of the cooperative (e.g., a gain from the sale of an asset) cannot be distributed as a patronage refund. Patronage refunds can be in the form of cash or equity. When patronage is distributed in the form of equity, it is typically called a *retained patronage distribution* or a *stock patronage refund.*

Qualified Versus Nonqualified

The term *qualified* refers to whether the patronage distribution is tax-deductible to the cooperative in the current year. Cash patronage refunds are always qualified. Cash patronage is therefore tax-deductible to the cooperative and taxable income to the patron in the year during which it is distributed. Retained patronage refunds (stock patronage refunds) can be either qualified or nonqualified. A qualified stock patronage refund is tax-deductible to the cooperative and taxable to the member in the year during which it is distributed. A nonqualified stock patronage refund is tax-deductible to the cooperative and taxable to the member in the year during which it is redeemed. When issuing nonqualified retained patronage, the cooperative pays taxed on that income in the year of distribution and then receives a tax deduction in a later year when the equity is redeemed. The taxation is ultimately transferred to the patron, but the timing is shifted to match the cash payment rather than the stock distribution.

In order for a cooperative to make a qualified patronage distribution (cash or stock):

1. It must be made within eight months from the end of tax year during which the income occurred.
2. The member must receive written notice of allocation.
3. At least 20% of the total distribution must be in the form of cash.
4. The member must consent to include the patronage on their tax return as ordinary income.

The rationale for the 20% cash rule is that the member should have enough cash to pay the taxes on the stock patronage received. Many producers have effective tax rates above 20% and would still be in a negative cash flow position if they received a stock patronage refund made up of 20% cash and 80% qualified. For this reason, most agricultural cooperatives try to pay in more than 20% cash. Note that the 20% cash requirement applies to qualified stock distributions. There is no cash requirement associated with a nonqualified stock patronage refund. The patron consent requirement is typically satisfied by having the patron sign an agreement as part of the membership application or specified as part of endorsing and cashing a qualified check.

If the cooperative does not meet any of the requirements for a qualified distribution or chooses not to classify the distribution as qualified, the distribution becomes nonqualified. Because cash patronage is always qualified, it is only the stock portion of the patronage refund that could be structured as nonqualified. Nonqualified patronage distributions cannot be deducted in the year during which they are issued, but the cooperative can deduct the redemption of nonqualified stock. This nonqualified redemption becomes taxable income for the patron during the redemption year.

Historically, most agricultural cooperatives distributed patronage in combinations of cash and qualified stock. This stemmed from the historic situation in which producers faced low tax rates while the corporate tax rate was much higher. This made it logical to shift the tax to the patron's lower tax rate immediately rather than park the tax payment with the cooperative until the time of equity retirement. The Tax Cuts and Jobs Act of 2017 substantially reduced the corporate income tax from a maximum of 37% to a flat rate of 21%. This resulted in cooperatives having lower tax rates than those of most of their patrons. I have done substantial research and simulations involving represented cooperatives and found research illustrating the fact that nonqualified distributions maximize the members' return from the cooperative. At the request of the National Council of Farmer Cooperatives, our academic group of cooperative specialists conducted national workshops and webinars on this subject. This effort led to a national award from the American Agricultural Economics Association for Outstanding Group Extension Program. Currently, we are seeing greater interest in nonqualified distributions, and more agricultural cooperatives are shifting to nonqualified.

[illegible] Comparison of Tax and Timing of Tax on Cooperative and Patron From Qualified and Nonqualified Stock Patronage

	Cooperative		**Patron**	
	Qualified	**Nonqualified**	**Qualified**	**Nonqualified**
Income before patronage	$100	$100		
Cash patronage	$20	$20	$20	$20
Qualified stock patronage	$80	$80	$80	$80
Taxable income	$0	$80	$100	$20
Tax	$0	$16.80	$21	$4.20
Redemption year				
Redemption	$80	$80	$80	$80

Table 15.1 Comparison of Tax and Timing of Tax on Cooperative and Patron From Qualified and Nonqualified Stock Patronage

Taxable income (+) or deduction (–)	$0	–$80	$0	+$80
Tax effect	$0	–$16.80	$0	$16.20
Final tax paid	$0	$0	$21	$21

For simplicity, a 21% tax rate is assumed for both the cooperative and patron.

Note that there is no tax effect for either the cooperative or the patron when qualified stock is redeemed since the member already paid the tax and the cooperative already received the tax deduction when the qualified stock was issued.

Taxation of Per-Unit Retains

A *per-unit retain* is a deduction of an amount of equity from a unit of commodity marketed. Some marketing cooperatives operate as pooling cooperatives. All the commodities delivered by the patrons are pooled and then processed, packaged, and marketed by the cooperative. The patrons often receive an intermediate payment to cover their production expense and then receive a final distribution when all the commodities are sold and the pool is closed. Although one could think of the intermediate payment as a commodity payment and the final payment as patronage, there is actually no dividing line because there is no defined price for the commodity. Pooling cooperatives therefore do not have patronage and cannot make retained patronage distributions (stock patronage). The per-unit retain is an alternative means for the cooperative to create equity. Per-unit retains differ from retained patronage in that they are based solely on the units of commodity handled and not on profits. Per-unit retains can be qualified or nonqualified and revolved under the same types of systems as retained patronage. The 20% cash patronage rule does not apply to a qualified per-unit retain.

When cooperatives make payments (as opposed to deducting equity) to the members under the per-unit retain system, it is termed a *per-unit retain paid in money* (PURPIM). So, instead of the cooperative purchasing commodities from the member at a given price, the member simply receives a PURPIM payment. The PURPIM structure was once unique to specialty crop pooling payments. In recent years, tax deductions such as the Domestic Production Activities Deduction and Section 199A have been available to marketing cooperatives. These tax deductions are basically calculated on the cooperatives' gross margin after producer payments to purchase commodities. PURPIM payments are not included in the income calculation. This created an advantage for a marketing cooperative to structure producer commodity payments as PURPIM because the deduction was based on a larger gross income. The Section 199A deduction, which is discussed in greater detail in a later section, made the PURPIM structure much more prevalent. The availability of the tax deduction has also increased interest in nonqualified patronage since the cooperative can offset the cooperative-level tax with the tax deduction.

Section 521 Cooperatives

Subchapter T also contains provisions for a more restrictive form of cooperative, designated a Section 521 cooperative. These cooperatives are sometimes termed *exempt* cooperatives, but this term is misleading because they are exempt only from some forms of taxation. In order to qualify for Section 521, a cooperative must be a farmer or fruit grower cooperative operating on a cooperative basis and involved in marketing farm products and/or providing farm supplies and equipment, in addition to meeting the following requirements:

1. A total of 85% of the capital stock must be owned by producers who market or purchase from the cooperative in the current year (85% stock owned by active producers).
2. Dividends on capital stock must be limited to 8%.
3. The cooperative must conduct at least 50% of its marketing business with members and 50% of supply business with members, and total marketing and supply business with nonmembers cannot exceed 15%.
4. The cooperative must treat nonmembers the same as members for patronage, pricing, and pool payments (i.e., it must pay patronage to nonmembers).

A Section 521 cooperative must also maintain records of patronage and equity and not have excessive levels of financial reserves (unallocated equity). Section 521 cooperatives are often exempt from U.S. Securities and Exchange Commission security registration laws. For that reason, many closed-membership, new-generation cooperatives choose to organize as Section 521 cooperatives. Under the closed-cooperative model, this type of cooperative had no nonmember business, and 100% of its members did business every year, so it had difficulties meeting the Section 521 restrictions. It is interesting to note that many of the Section 521 requirements appear to relate to the Rochdale principles.

The tax advantages of Section 521 cooperatives is that while they have to pay patronage on nonmember business, they can deduct those patronage payments along with all qualified patronage and per-unit retain payments to members. In essence, a Section 521 cooperative operates and is taxed as if all its business is in member business. Additionally, a Section 521 cooperative can deduct dividends paid on capital stock, whereas those dividends are limited to 8%.

Nonmember Business and Unallocated Retained Earnings

Cooperatives are taxed at the ordinary corporate rate on their taxable income, which is calculated after cash patronage, qualified retained patronage, and qualified per-unit retain payments. All cooperatives except Section 521 cooperatives must include nonmember income in their taxable income. All cooperatives, including Section 521 cooperatives, must also include income from rents, investment revenues, gain on sale of capital assets, and income from sales to the federal government as taxable income. For that reason, even Section 521 cooperatives often have some taxable income. Cooperatives generally retain nonmember and nonpatronage income as unallocated retained earnings. Because of the tax effects, is it actually the after-tax portion of those profits that is retained. If the cooperative was in the 21% tax bracket and received $100 of nonpatronage income, it would pay $21 in tax and retain the after-tax portion of $79.

Cooperatives also have the option of retaining a portion of member profits as unallocated retained earnings. Because that income is not distributed as qualified retained patronage, it is not tax-deductible, and the cooperative actually retains the after-tax portion. The availability of the Section 199A tax deduction has allowed cooperatives to offset that tax effect and led some cooperatives to retain more member profits as unallocated retained earnings. Cooperative scholars would argue that a better approach would be to retain those profits as nonqualified retained patronage and use the tax deduction in the same manner. By retaining the profits as nonqualified, the cooperative will receive a deduction in the equity retirement year. Nonqualified stock essentially allows the cooperative to bank the tax deduction for a future year.

Taxation of Federated Cooperatives

Federated cooperatives are taxed in the same fashion as the local cooperative. When a federated regional cooperative makes a cash or qualified stock patronage refund to the local cooperative, that patronage becomes part of the local cooperative taxable income. The local cooperative must distribute that income within eight-and-a-half months to its local patrons in a tax-deductible form (e.g., cash, qualified stock, or qualified per-unit retain distribution); otherwise, it must pay tax on that income. The treatment of dividends on capital stock is more complex. A Section 521 cooperative is allowed to deduct 80% of the capital stock dividends it receives from other cooperatives. A Section 521 cooperative is allowed to deduct all of the capital stock dividends it receives from other cooperatives as long as it distributes them to its patrons.

A look-through procedure is used to determine the Section 521 status of a regional cooperative. The cooperative must meet the Section 521 requirements regarding its producer members, and if it has cooperatives as members, the IRS will look through to determine whether those cooperatives meet the Section 521 provisions.

Tax Credits and Deductions

As discussed, marketing cooperatives have been able to use the Section 199A tax deduction in recent years, Congress has occasionally implemented other tax credits and deductions such as investment tax credits for investment in depreciable property or accelerated depreciation tax deductions. Those tax credits and deductions are often not useful to cooperatives since they severely reduce or eliminate their taxable income by making patronage payments. Tax rules often do not allow a cooperative to carry unused tax credits backward or forward to other tax years as other businesses do. Cooperatives are sometimes allowed to allocate unused tax credits to their members similar to the way they allocate patronage refunds. Under the current Section 199A, producers who market through a cooperative face a tax penalty. This penalty is the reduction in an otherwise available tax credit. The cooperative can allocate all of a portion of its Section 199A tax deduction to its patrons. Most marketing cooperatives attempt to allocate sufficient tax deduction to keep their members equivalent with producers marketing through noncooperatives and retain the rest to offset the taxation from issuing nonqualified stock.

Other Taxes

Cooperatives are usually subject to all other taxes on the same basis of other businesses. Cooperatives pay sales, payroll, license, property, and excise tax. In some states, cooperatives are exempt from corporate franchise taxes, which are taxes on the net worth of corporations. States with income tax generally follow the basic provisions of federal tax laws in taxing cooperatives.

Taxation of Other Cooperatives

The tax rules discussed apply to farmer-owned cooperatives. Other cooperative businesses may fall under different tax regulations. Section 501 of the IRC deals with not-for-profit corporations. Religious, charitable, and civic institutions owned and operated for the benefit of their members fall under Section 501. Some cooperatives, including some credit unions, mutual insurance agencies, and rural electric and telephone cooperatives, are able to be taxed under Section 501. CoBank and most other branches of the Farm Credit System operate on a patronage basis and are taxed under Subchapter T. Consumer cooperatives are taxed under Subchapter T, and since Section 521 cooperatives must be owned by agricultural producers, there is no such thing as a Section 521 consumer cooperative. Consumer cooperatives deduct patronage refunds in a similar fashion to farmer-owned cooperatives. Patronage refunds related to purchases of personal living or family items are not taxable income to the patrons. The logic is that unlike an agricultural cooperative, where the cooperative activities are an extension of the farm business, the patrons of consumer cooperatives are purchasing with after-tax income. Consumer cooperatives are the only type of cooperative that can exclude patronage refunds from taxable income.

Tax Examples

Tables 15.2 and 15.3 provide some examples of tax calculations for cooperatives. Although at first glance the calculations appear complicated, there are some simple steps to take to demystify them. The cooperative's local income in these scenarios is $11,000. Regional patronage is only assumed in the third scenario which raises the total income to $12,000. In the other scenarios, no regional patronage is assumed so local income and total income are identical. For the non–Section 521 cooperative, we can add the actions that create deductions for the cooperative (cash patronage, qualified stock patronage, and redemption of nonqualified stock patronage) to determine the total amount of deductions. This allows us to calculate the cooperative's taxable income and tax payment. We can then add up the cooperative's cash outlays (cash patronage, equity redemption payments, stock dividends, and taxes). Subtracting the total cash outlay from the original total income yields the cooperative's cash flow. The member's taxable income is calculated by adding the payments that are taxable income for the member (cash patronage, qualified stock patronage, and redemption of previously issued nonqualified stock). Dividends paid on stock do not create a deduction for the cooperative but do create taxable income for the stockholder. We separate that tax burden from the member since preferred stock in a cooperative can be owned by nonmembers. Note that the total taxable income for the cooperative plus the member and stockholder always adds up to the original total income. You might note that in these examples, the cooperative chose not to change their cash and stock patronage when they received regional patronage.

Table 15.2 illustrates similar situations for a cooperative that qualifies for Section 521. The calculations are identical except that patronage is calculated on the total (member and nonmember) profits, and the dividend paid on stock continues to create a cash outlay but now also creates a tax deduction.

Table 15.2 Tax Examples for Non–Section 521 Cooperatives

	Non–Section 521 Qualified	Non–Section 521 Nonqualified	Non–Section 521 Regional Patronage	Non–Section 521 Redeeming Nonqualified
Member-based profits	$10,000	$10,000	$10,000	$10,000
Nonmember-based profits	$1,000	$1,000	$1,000	$1,000
Regional patronage	$0	$0	$1,000	$0
Total potential taxable income	$11,000	$11,000	$12,000	$11,000
Stock dividend	$0	$0	$0	
Cash patronage	$5000	$5000	$5,000	$5,000
Qualified stock patronage	$5000		$5,000	$5,000
Non-qualified stock patronage		$5000	$0	
Non-qualified redeemed	$0	$0	$0	$500
Total deductions	$10,000	$5,000	$10,000	$10,500
Taxable income (A)	$1000	$6,000	$2,000	$500
Tax (25%)	$250	$1,500	$500	($125)
Cash payments (not considering depreciation and other effects)	$5250	$6,500	$5,500	$5,375
Cash flow	$5750	$4,500	$6,500	$5,625
Member taxable income (B)	$10,000	$5,000	$10,000	$10,500
Total taxable income (A+B)	$11,000	$11,000	$12,000	$11,000

Tax Examples for Section 521 Cooperative and One Complex Example With a Non–Section 521 Cooperative

	Section 521 Cooperative With Qualified	**Section 521 Cooperative With Nonqualified**	**Section 521 Cooperative With Qualified Paying Dividend**	**Non–Section 521 Cooperative: Regional Patronage, Mix of Qualified and Nonqualified, Stock Dividend, and Redeeming Nonqualified**
Member-based profits	$10,000	$10,000	$10,000	$10,000
Nonmember-based profits	$1,000	$1,000	$1,000	$1,000
Regional patronage	$0	$0	$0	$1,000
Total potential taxable income	$11,000	$11,000	$10,000	$12,000
Stock dividend	$0	$0	$1,000	$1,000
Cash patronage	$5,500	$5,500	$5,500	$5,000
Qualified stock patronage	$5,500	$0	$5,500	$2,500
Nonqualified stock patronage	$0	$5,500	$0	$2,500
Nonqualified redeemed	$0	$0	$0	$500
Total deductions	$11,000	$5,500	$12,000	$8,000
Taxable income (A)	$0	$5,500	($1,000)	$4,000
Tax (25%)	$0	$1,375	($250)	$1,000
Cash payments (not considering depreciation and other effects)	$5,500	$6,875	$6,750	$7,500
Cash flow	$5,500	$4,1,25	$4,250	$4,500

Table 15.2 Tax Examples for Section 521 Cooperative and One Complex Example With a Non–Section 521 Cooperative

Member taxable income (B)	$10,000 (plus $1,000 nonmember)	$5,000 (plus $500 nonmember)	$10,000 (plus $1,000 nonmember plus $1,000 stockholder)	$8,000
Total taxable income (A + B)	$11,000	$11,000	$11,000	$12,000

Summary

The principal of cooperative taxation under Subchapter T is that cooperative net income is generally taxed at either the cooperative level or the patron level but not both. Subchapter T allows cooperatives to deduct certain types of patronage refunds from their taxable income, which is the vehicle that accomplishes single taxation. Cooperatives are taxed on any remaining income, including nonmember income, nonpatronage income, and member income not distributed as patronage, at the regular corporate rate.

A patronage refund that is deductible to the cooperative and taxable to the member is termed a *qualified* refund. Cash patronage is always qualified, although it is seldom termed this way. Retained patronage refunds (stock patronage) can be either qualified or nonqualified. Cooperatives can deduct nonqualified stock patronage when they redeem it. In the case of nonqualified stock patronage, the cooperative pays tax on that income in the distribution year but eventually receives a deduction when it is redeemed. The ultimate effect is that the member pays taxes on the profits represented by the nonqualified distribution, but the timing of the tax effect differs from that of qualified stock, for which the patron is taxed upon receiving it.

Section 521 cooperatives are a more restrictive form of cooperatives and must meet certain requirements. The major tax advantage of Section 521 is that those cooperatives can deduct patronage paid to nonmembers and dividends paid on capital stock. Federated cooperatives are taxed similar to local cooperatives. When a local cooperative receives cash or qualified stock patronage from a federated regional cooperative, they must pass it on to their patrons in a tax-deductible form with eight-and-a-half months or pay taxes on that income.

Cooperatives qualify for some tax deductions and credits, similar to other businesses. These tax deductions and credits may not be useful to cooperatives if they are distributing their income as qualified patronage since they will have very low taxable income. In some cases, cooperative can allocate unused tax deductions and credits to their members in a similar manner as patronage. In recent years, agricultural marketing cooperatives have made extensive use of the Section 199A deduction. They have pooled that deduction from all of their members' production activities and then retained a portion at the cooperative level while distributing the remainder to members. That has created another channel for member benefit and another complex balancing act for the board of directors.

CHAPTER 16

Strategies for Supply Cooperatives

Introduction and Learning Objectives

This chapter focuses on the strategies and activities of what are usually called *farm supply cooperatives.* Traditionally, supply cooperatives concentrated on purchasing farm inputs to be resold to their members. Many farm supply cooperatives also offer services such as seed cleaning, fertilizer application, and crop scouting. Some scholars would expand that definition of farm supply cooperatives to include shared service cooperatives such as artificial insemination, utilities (e.g., electricity, telephone, and water), and credit. In this chapter, we concentrate primarily on traditional farm supply cooperatives providing farm inputs and related services. We should also note that there are many wholesale cooperatives serving small businesses. Those cooperatives could be described as purchasing cooperatives (from the perspective of the business owner-members) or as supply cooperatives. Retail cooperatives such as cooperative groceries and food markets are also in some sense supply cooperatives. Those cooperative sectors operate in different environments and do not face the same situations and strategies discussed in this chapter.

Rational for Farm Supply Cooperatives

As with many agricultural cooperatives, the basic rationale of farm supply cooperatives is to provide economy of scale in sourcing and, in some cases, applying agricultural inputs. These scale economies are particularly obvious with bulk commodities such as fertilizer, bulk chemicals, and petroleum. Farm supply cooperatives tend to have a very broad product line and diverse membership. They may at times offer contracts or prepricing on a few commodities, but most sales are on an open-purchase arrangement. These cooperatives operate very similarly to investor-owned retail supply firms that purchase, display, price, and sell supplies. Farm supply cooperatives may deliver fuel, feed, and fertilizer to their members. They depend on a favorable marketing mix of product lines, prices, advertising, services, and potential patronage refunds to attract customers. Members and nonmembers alike can freely shift purchases to another firm. These cooperatives exemplify the Edwin G. Nourse philosophy of being a competitive yardstick—that is, they provide a comparison for the fair price of agricultural inputs.

Product-Line Considerations

Because farm supply cooperatives have a diverse membership and broad product line, they need to determine the appropriate mix of goods and services that they offer members. Because they are member-owned, decisions about the product line and inventory can become more political and controversial than those in an investor-owned firm. Any firm considering an expansion of its product line should consider such issues as spreading of costs and risks over more products and seasonal demands in terms of labor and plant capacity. They must also avoid spreading management, financial, and other resources too thin. Farm supply cooperatives should also determine whether it will be advantageous for them to expand their product line. Since not all members use all products, there is also the danger of the overall membership subsidizing a subset of the members who use a particular product. Logistically, a cooperative can operate most efficiently with a single-product or specialized product line. On the other hand, wider product and service lines are convenient to patrons. Improved economies can sometimes be realized if additional product lines share some of the same fixed costs. Further, multiple-product cooperatives may be better positioned to discourage competitors from entering their trade territory and provide services that would not otherwise be available. Finally, board members and management benefit by exposure to a wider variety of opportunities and contacts.

The relative importance of these issues varies, depending on the circumstances. The following questions should be considered when evaluating a new product or service:

1. Will a new product or service duplicate adequate services already provided by another cooperative in the geographic area? If so, members may be better off by working through the other cooperative to avoid unnecessary duplication.
2. Is there reasonable assurance that net income generated will repay the investment and provide adequate returns to patrons?
3. Will the expanded product line fulfill a genuine need of members, one that they are willing to pay for and patronize?
4. Will the product complement the existing management and physical plant? This should be considered in calculating return on investment.

Evaluating these factors is a responsibility of the manager and board. Managers typically make tactical product line decisions, such as which animal health products to stock, whereas the board makes strategic product line decisions, such as whether to carry hardware or work apparel. Appropriate answers will be different in different situations and will change with time. The board must continually look at the feasibility of alternative product lines as it strategically positions the cooperative in a rapidly changing economic environment.

Competition Between Cooperatives

While cooperation between cooperatives is an aspect of the International Cooperative Alliance principles, farm supply cooperatives also compete with other cooperatives. They also compete vigorously with noncooperative firms in their trade territory. Members often favor having multiple cooperatives competing for their business. They believe that this competition eliminates inefficiencies and keeps management honest. The downside of cooperatives competing is that it leads to overcapacity and inefficiency. Members ultimately must bear the cost of duplication, such as delivery trucks from competing cooperatives passing each other as they travel over overlapping trade territories. The argument for multiple cooperatives competing is that it gives members a direct comparison of price and service. A counterargument is that members can judge the performance of the firm through its annual results and should favor the most efficient structure of nonduplication.

New Products and Services

Cooperative literature is rich in the concept that cooperatives exist to serve their members. Cooperatives are member-based organizations. They need to be responsive to members' needs and wishes. But members have a propensity for prevailing on their board for an ever-widening assortment of goods and services, often expecting to receive the services free of charge or at a subsidized rate. When this elevates the cooperative's cost structure, these same members abandon their cooperative to patronize another business that offers favorable prices because it has focused on a narrow, high-volume line of products and services. On the other hand, a new product or service can often be carried by an existing assembly/distribution system at minimal marginal cost. Such additions may be beneficial, particularly if the expanded product line is priced to pay at least the marginal costs associated with it.

Nonmember Business

Farm supply cooperatives often have a higher portion of nonmember business than marketing cooperatives. Profits from nonmember business are typically retained as unallocated retained earnings and thus provide a reserve fund for the cooperative. Nonmember business can therefore be advantageous since the profits indirectly benefit the members. Nonmember business may also allow the cooperative to lower costs by increasing volume or use labor and facilities more efficiently. Nonmembers may also decide to eventually join the cooperative in order to obtain patronage, so nonmember business can be a mechanism to increase membership. The downside of nonmember business is that the risks are ultimately borne by the members. As an example, in times of expanded oil field drilling activity, farm supply cooperatives can obtain attractive margins by selling fuel to those firms. However, if the oil field economy suffers a downturn and those firms go bankrupt or dissolve, the agricultural members will ultimately absorb the loss from uncollectable accounts.

Pricing Policies

Price is an important part of the marketing mix for any firm that sells products and services. In theory, marketing firms should consider the elasticity (price sensitivity) of their consumers. Cooperative principles suggest that farm supply cooperatives should price at the market level. Because they have a broad product lines, it is often difficult to determine a clear prevailing market price. Strategies for pricing include premium pricing, favorable pricing, and cost-plus pricing.

Premium pricing refers to setting prices to generate margins as large as competitive pressures will allow. In theory, a cooperative using a premium price strategy should have higher profits that can be returned through patronage. In practice, members often object to strategies that involve the highest prices possible. Premium pricing also leads to lower volume, which may not allow the firm to spread fixed costs across as many units. Favorable pricing (pricing supplies below market prices) passes the benefits immediately to patrons in the form of low prices for supplies and high prices for products. Cooperative members often push the firm to offer products at favorable prices. The downside of favorable prices (which has been discussed in previous chapters) is that other firms match the prices (benefiting nonmembers) and members do not recognize the price benefits as coming from the cooperative. Because of the difficulties in determining market prices and customer price sensitivity, many cooperatives use a cost-plus pricing strategy in which they try to maintain a given percentage markup or a given profit margin for each product or category of products.

Differential Pricing

Differential pricing means offering different prices for different sales situations. Differential pricing is practiced in several industries. For example, airlines use peak versus off-peak pricing, and utilities distinguish between industrial and residential rates. Ride-sharing companies such as Uber and Lyft price every transaction at a different rate depending on the supply of drivers and number of customers seeking rides. Cooperatives may offer discounts used by other businesses, such as cash and seasonal discounts. These discounts have few or no unique features among cooperatives and are generally accepted by members. Therefore, we will not discuss them further. In this section, we focus instead on the contrast between average-cost pricing, often referred to as *equal pricing*; differential prices based on volume per transaction, often referred to as *equitable prices*; and differences in demand.

Under average-cost pricing, every cooperative member faces the same price for a product or service. At one time, the use of average-cost pricing by agricultural cooperatives was the dominant approach. This policy was appropriate as long as the size and enterprise combinations of members were relatively uniform. But the changing size distribution of cooperative patrons, some with small, part-time farmers and some from very large farms, led to different costs associated with servicing different-sized operations. Cooperatives also recognized the importance of maintaining the large-volume customers, who had the greatest opportunities to bypass the cooperative. As cooperatives found that they must aggressively compete for the business of large-scale members, differential pricing became more prevalent. Under a differential pricing strategy, different segments of the cooperative membership would face different pricing. The goal of differential pricing is to attract additional customers who would not otherwise use the cooperative with more favorable pricing while maintaining higher margins with other segments who are less price-sensitive. Even though it can make good business sense, differential pricing can be controversial in a cooperative.

Opposition to Differential Pricing

There are several common objections to differential pricing in a farm supply cooperative. These include the following:

1. Members with small operations feel that they are always being discriminated against. They joined the cooperative to offset market power by large firms. They feel that differential pricing is not fair. Small-volume patrons believe that large-volume patrons should not receive both a price break and a patronage refund.
2. Cooperatives are a people-oriented business. Treating different groups of members differently is therefore against the spirit of the cooperative structure.
3. The erroneous assertion exists that uniform pricing is a basic cooperative principle.
4. Differential pricing is a nuisance. The additional member education and recordkeeping are not worth the cost.

These ideas probably stem in part from a commitment to the one-member-one-vote principle and from members' feeling of frustration in dealing with market powers on multiple fronts. Members join cooperatives to obtain economies of size unavailable to them as individuals, and differential pricing appears to them to defeat that purpose.

Objections to differential pricing reflect a misunderstanding and erroneous beliefs. Equal treatment, or uniform pricing, is not a legal requirement. Different prices can be used as long as they are based on costs or competitive pressures. Equal treatment is not a cooperative principle. In fact, strict adherence to equal prices violates the principle of business at cost because members who are less costly to service should probably be charged less.

Equal-Margin Pricing

Equal-margin pricing reflects differences in the cost of providing a service to different classes of patronage so that the net margin—price minus costs—of each category is equal. For example, if it costs $0.04 per gallon less to make a 2,000-gallon delivery of petroleum products than to make a 200-gallon delivery, a $0.04-per-gallon discount is appropriate for the larger delivery. Equal-margin pricing reflects the fact that equal pricing is not always equitable pricing. Equitable prices should accurately reflect the costs of resources employed. For example, the quantity discount schedule for a petroleum cooperative was $0.05 per gallon for 2,000 to 4,000 gallons, $0.08 per gallon for 4,000 to 8,000, and $0.12 per gallon for tanker loads of 8,200-gallon deliveries. With discounts such as these, members may be motivated to install larger storage tanks and thereby reduce the cooperative's delivery costs.

A major concern for cooperatives is that equal-margin prices should be based on cost. Therefore, management must periodically evaluate relevant costs to ensure that price differences accurately reflect associated costs. Otherwise, one class or category of patronage may end up subsidizing another. For example, if the cash discount for early payment is 10% when the actual savings from cash payment is 15%, cash patronage is subsidizing credit patronage.

Demand Differences

Different categories of patrons have different demands for products and services, creating the potential for the cooperative to structure differential prices based on demand. For example, large producers who have the ability to go around the cooperative might be attracted by lower fertilizer prices, whereas lifestyle farmers may be less concerned about the price. Demand difference can also relate to location. For example, suppose that patrons are scattered over the market area of a fertilizer cooperative. The net farm price to members is the posted price plus the farmer's pickup costs. Because these costs depend on distance, producers located near the cooperative face a lower net price relative to those farther out. The more closely located producers will also probably face a longer distance to source from a competitor. The producers located close to the cooperative might therefore be less sensitive to prices than those located on the edge of trade territory.

Price differences based on demand rather than cost place cooperatives in a dilemma. Even though all patrons may benefit when the cooperative uses differential pricing to increase volume and lower average costs, demand-based price differences are not based on the business-at-cost principle.

When Are Differential Prices Not Justified?

Differential pricing is not justified when the complexity of the pricing system outweighs the potential benefits. Cooperatives should also not use differential pricing if the majority of members oppose it. Even if large-volume members leave, the cooperative will be obliged to continue with single prices as long as the majority of members insist on them. However, it must be remembered that the remaining members will eventually end up with higher costs. Hopefully, an educational program about the realities of the competitive environment can change opposing members' feelings.

Legal Implications

The Robinson-Patman Act of 1936 provides that differential pricing is legal if the differences are based on costs or differential pricing is used as a defensive move to meet the prices of competitors. Concern about making a legal justification for different prices is unnecessary as long as differential patronage refunds ultimately result in uniform prices and business at cost.

Netting

Netting is a policy of establishing prices so that the losses of one division are absorbed by the net income of another division. In effect, a cooperative pools income and losses across its various divisions. Cooperatives use netting in cases where they need to spread risk, absorb developmental costs associated with a new product, or extend market area. Netting also avoids the complicated decisions of allocating losses to members. For example, if the fertilizer department experienced a loss, few cooperatives would want to write down the stock of members who purchased fertilizer when the overall cooperative was profitable.

Other less noble examples also exist. In one case, the board of a local supply cooperative voted to keep an automobile service bay open even though they knew it was losing money. The problem with this action was that members not using the service subsidized those that did. Netting often occurs because cost analysis is not done carefully enough for management to realize what is occurring. Netting should be avoided as a deliberate policy because it violates the business-at-cost principle.

Current Assets

Farm supply cooperatives typically have significant investments in both inventories and accounts receivable. In many cases, the seasonal line of credit exceeds the amount of the cooperative's long-term loan. Members of farm supply cooperatives are typically interested in open purchase arrangements, in which they have no commitment to purchase from the cooperative. Farm supply cooperatives often end up with substantial investments in inventories. They also face substantial risk if the price of the product declines. For example, if the price of fertilizer decreases, members will source from another outlet unless the cooperative matches the current market price. The resulting price decrease can be much larger than the cooperative's margin, resulting in a substantial loss to the firm. For this reason, some cooperatives push members to prepay or commit to purchases before they purchase the inventory. Members often have a knee-jerk reaction against prepurchase policies. However, having the cooperative absorb all inventory price risk is not fair to the members who are not purchasing the particular input.

Cooperatives also have a large investment in accounts receivable. Short-term credit is generally intended as a convenience to members. For example, it would be inefficient during a hectic planting period for the member to write a check for every load of fertilizer they picked up. Accounts receivable pose a significant risk to the cooperative. Some members will use the cooperative for credit because they have exhausted all other credit sources. Accounts receivable create significant costs in recordkeeping, monitoring, and interest. It is therefore essential for the manager to monitor and collect accounts receivable so that one subset of the membership is not benefiting at the expense of the overall membership. Because the customer is also an owner, enforcing the credit policy and collection of past-due accounts can be a delicate process. Many cooperatives are also reluctant to ask their member-owners for financial statements. This can make it difficult for the cooperative manager to assess the creditworthiness of the customers. Some cooperatives that have historically had problems managing accounts receivable enroll in services that will take on the credit process for a fee. In these cases, the outside firms have the credit risk. The firms that take on this process typically require full financial statements from the members and use other methods to judge creditworthiness. This may result in some members being unable to charge through the cooperative. In essence, the outside firm simply has the discipline to perform the credit management procedures that the cooperative was unable or unwilling to perform.

Summary

Supply cooperatives face unique challenges and opportunities in developing and adopting strong marketing programs. The purpose of this chapter was to explore distinctive marketing issues of supply cooperatives. Particular attention was devoted to a cooperative's product and pricing strategies.

Cooperatives must evaluate the tradeoff between the benefits of greater uniformity among its members and the specialization associated with a narrow product line, on the one hand, with the benefits of sharing fixed costs and one-stop service associated with a broad product line, on the other. Agricultural cooperatives in many areas need to evaluate overlapping trade areas and duplicated facilities when developing their marketing strategies. Expanding nonmember business is a strategy to help cover fixed costs and capitalize on the cooperative's image and provide synergies with its other activities.

Pricing strategies, involving price-level and volume discounts and premiums, dictate how benefits are distributed. Premium pricing, or the use of prevailing market prices or relatively high margins, provides cash flow for the cooperative, protects against losses, and avoids the free-rider problem. Favorable prices, on the other hand, pass benefits immediately to patrons, encourage increased patronage, and may affect the price structure of the relevant market.

Equal pricing, or charging the same price to all members, has been a traditional cooperative practice. The appearance of greater heterogeneity between firms and farms and increased competitive pressures have led to differential- or equal-margin pricing, which entails offering discounts and premiums that reflect benefits from a large volume or differences in demand. This sometimes controversial practice, if properly administered, benefits all members, is legal, and is in harmony with cooperative principles. The appropriateness of netting, may be appropriate when spreading risks and absorbing developmental costs; otherwise, it is questionable.

CHAPTER 17

Marketing Strategies for Marketing Cooperatives

Introduction and Learning Objectives

Marketing cooperatives are primarily unique to agriculture and aquaculture. They operate as an extension of the farm firm, marketing their members' commodities. In many cases, marketing cooperatives also grade, store, process, package, and transport commodities. Some marketing cooperatives also provide farm supplies and can be classified as farm supply and marketing cooperatives. Marketing cooperatives and the marketing activities of diversified cooperatives involve unique issues and strategies that will be discussed in this chapter.

Commodity Specialization

While most supply cooperatives carry a broad line of supplies, marketing cooperatives often handle only one or a closely related family of commodities. Farmers may be selling only one farm product and require marketing services just for that single commodity. In addition, the required marketing functions may vary between commodities. For example, milk marketing is considerably different from grain or cotton marketing. An established cooperative may market additional commodities from its members as a service to them, add commodities whose products can be marketed in its established marketing channel, or add new products as a complement to existing product lines. Cherry Central, for example, added apples and fresh produce. Many Oklahoma grain cooperatives that historically handled winter wheat now also store and market summer crops such as corn, soybeans, and milo.

Acquisition Strategies With Members

Marketing cooperatives use different strategies to acquire commodities from members. These strategies vary both with the level of commitment or integration between the cooperative and the member and the physical operations performed by the cooperative. Marketing cooperatives can be structured to only provide market information (in which case the cooperative is not actually acquiring the commodity), as a bargaining cooperative establishing the price and terms of sale, or purchase commodities through cash on delivery, auction, delayed payment, or a marketing pool. The characteristics of these structures are discussed in the following sections.

Information-Sharing Cooperatives

Some marketing cooperatives exist primarily to provide market information to their members. This form of cooperative was originally developed in California, and the first information-sharing cooperative was formed by lettuce growers. Melon, kiwifruit, table grapes, stone fruit, and mushroom producers have followed their example. The birth of these cooperatives was triggered by producers forced to market their products in very unstable markets with a paucity of information. These markets are characterized by a relatively small number of producers controlling a major portion of U.S. production of highly perishable products with relatively inelastic demands.

Information-sharing cooperatives provide an organized framework to (1) summarize member information such as inventories, sales on those inventories, and projected production and sales and (2) hold periodic meetings during the sales season. These meetings have been characterized as loose, free-wheeling discussions regarding what prices should be in light of market conditions. After these meetings, members establish their own pricing and related marketing strategies. Information-sharing cooperatives do not engage in traditional marketing functions, such as title transfer or logistics. They do not even negotiate with buyers for better terms of trade. Many are operated by a contract management company and financed by membership dues or a check-off (per unit assessment on products marketed) plus meeting charges.

Bargaining

Bargaining cooperatives or associations are organizations of producers of one or more commodities that represent their members in negotiating with processors and other buyers for the best terms of trade, including price, in light of economic realities. Emphasis is placed on terms of trade rather than just on price, because other considerations such as payment provisions, delivery point, method of delivery, and quality premiums may more than offset a favorable price. Bargaining cooperatives dominate perishable products, such as fruits and vegetables, sugar beets, and milk. Most bargaining cooperatives negotiate for only one commodity. Fruit and vegetable bargaining cooperatives are concentrated in the Pacific states.

Bargaining Procedures

Producers sign individual contracts with the cooperative. In their purest form, bargaining cooperatives neither take title nor handle the product. They simply negotiate the contracts between growers and purchasers. Producers deliver their products directly to buyers who take title. Sometimes the cooperative may provide additional services, such as verifying grades and weights. As a bargaining chip, cooperatives also may arrange for improved efficiencies; for example, they schedule varieties planted to avoid harvest gluts, or they provide incentives for improved quality. In order to further enhance the associations' bargaining position, cooperatives may acquire processing and merchandising facilities of their own as an alternative outlet; they do this primarily as a warning to processors during negotiations but also to handle surpluses, which then are usually sold for lower-valued uses. The boards of directors, committees of members, or, in some cases, the entire membership negotiate regarding any issues. These groups generally consider relevant economic factors, such as production volume, carryover, international supply, competition, and general economic conditions. The views of members regarding the issues to be negotiated are sought regardless of which group makes the final selection.

Under Which Conditions Is Bargaining Likely to Be Successful?

Several elements are required for successful bargaining. The market structure characteristics that make control over supply most likely are (1) barriers to entry, (2) lack of close substitutes, (3) geographic concentration of production, and (4) a relatively small number of producers. Examples of barriers to entry are the high capital cost for dairy, geographic production restrictions of cling peaches, and the length of time to bring new pear trees into production. If terms of trade are improved over other alternative employments without barriers to entry, other producers will enter the market, increase aggregate supply, and cause havoc with bargaining efforts. If substitutes are available, consumers may switch to them and thus erode many benefits of bargaining, as has happened in the butter and cotton markets. Geographic concentration of relatively few growers facilitates organizational efforts. To capture control over supply, producers must also be organized. Creation of a bargaining cooperative representing all or at least a major portion of the producers immediately changes market structure. In the absence of a bargaining cooperative, processors face a large number of producers whom they could bid off against one another. When a bargaining cooperative exists, processors are confronted by an organization capable of restricting their access to supplies.

Because bargaining cooperatives do not buy and sell the commodity, they do not generate margins like marketing cooperatives. Revenue received by bargaining cooperatives comes from one or more of the following sources: check-offs, service charges paid by processors, annual dues, and membership fees. The check-offs are generally withheld from member payments at the processor level instead of being directly paid by producers. If annual dues are used, they are determined by a per-acre or per-unit basis and paid by the members. Membership fees are typically one-time charges made only when members join the cooperative.

While important in some commodities, bargaining cooperatives are not always successful. Gains may be modest, or the bargaining effort breaks down because of free riders or increased production. Processors encourage free riders by paying both members and nonmembers the negotiated price. If a bargaining cooperative creates more favorable prices, it can result in increased production from both members and nonmembers. The increased supply often decreases the bargaining position of the cooperative.

Marketing Orders

Although marketing orders (MOs) are not necessarily linked to a cooperative, many bargaining and marketing cooperatives operate in conjunction with federal MOs. MOs are institutions enabled by government legislation that farmers may use to promote collective orderly marketing of selected commodities. MOs can control quantity, quality, and/or marketing activities. The primary characteristic of an MO is that when the majority of farmers approve one, all farmers and processors must abide by its rules and regulations. Referendum votes are held periodically (e.g., every six years) to determine whether two-thirds of the producers are still in favor of the order. MOs are unique to agriculture. Cooperatives play an important role in their creation and maintenance. MOs often help cooperative bargaining efforts and reduce the free-rider problem.

Enabling legislation for MOs flows from the Agricultural Marketing Agreement Act of 1937. This act is enabling because it does not create MOs but rather makes their creation legal and establishes the institutional framework and procedures for their creation. Many agricultural commodity producers (e.g., milk, fruit, vegetable, and nut) can qualify. State and federal legislation allows MOs to stabilize erratic markets of selected farm commodities and promote reasonable commodity prices to provide consumers with an adequate supply of wholesome food. An MO can be created only after producers have requested it and shown that it is administratively and economically feasible. Hearings are held to determine whether more orderly marketing is needed, production areas (or consumption areas, in the case of milk) are independent, and an MO is not against the long-run interest of consumers. The conditions for an MO to be effective are similar to those for a successful bargaining cooperative. There must be (1) a perishable commodity, (2) some barriers to entry, (3) organized and knowledgeable leadership among producers, and (4) a geographic concentration of producers.

If conditions warrant an MO and a two-thirds majority of producers approve, marketing regulations are drawn up that control specified marketing functions. Only handlers (not producers) are regulated. They are required to submit reports to the administering agency regarding compliance with the order and product flows. MO regulations include factors such as quality, market flow for fruits and vegetables, minimum and blend prices for milk, blend prices for producers, packaging standards, research and promotion check-offs, elimination of unfair trade practices, and gathering and reporting of market information. MOs for milk are administered by an appointee of the secretary of agriculture. Orders for other commodities are administered by a committee elected by growers. Operational costs are covered by check-offs deducted from the prices received by farmers.

Several MOs are currently in operation. The United States is blanketed with 11 federal milk MOs that account for more than 80% of the country's milk. Fruit and vegetable MOs are scattered throughout the United States but concentrated in California. Several are also active in Florida, Texas, and Washington. A pecan MO was recently established in Oklahoma. Bargaining cooperatives have been relatively more successful when bargaining efforts are initiated under the umbrella of an MO. For example, dairy cooperatives have received considerable publicity for their success in obtaining MO premiums.

Marketing Cooperatives That Purchase Commodities

The next level in terms of member commitment with the marketing cooperative is the purchase and resell structure. Many marketing cooperative, including most that market grain and bulk commodities, purchase the commodity at a posted market price and then resell it in larger lots. In many cases, the cooperative also stores the commodity and manages transportation logistics. In some cases, it is also involved in processing the commodity. There are a number of structures for the purchase transaction.

Cash on Delivery

Cash on delivery (COD) means that cooperatives pay a cash price for and take title to products delivered by patrons. These products are then often comingled and sold in the market at the most advantageous price possible. Net income remaining after expenses flows to patrons through patronage refunds. Most grain and oilseed marketing cooperatives operate under cash at delivery.

There are several advantages of COD policies. Farmers know at once what they will get; uncertainty is removed. Many bulk commodity marketing cooperatives compete with investor-owned firms (IOFs) that purchase commodities for cash. That makes it difficult to convince members to accept structures that are more restrictive. Members have more individual control over their marketing strategies under the cash purchase structure.

There are also disadvantages. Cooperatives cannot exert their full marketing power and expertise when members control marketing decisions. For example, a wheat marketing cooperative cannot enter into agreements with a flour mill because it does not know how much wheat its members will decide to deliver. The cooperative takes on additional risks when they take title to commodities. For example, the cooperative can purchase wheat from the producer and then experience a large loss if the market price declines before the cooperative can sell it. The cooperative can transfer or offset many of these risks through appropriate risk management strategies, such as forward contracting and hedging. Another issue with cash purchase structures is that they require considerable working capital, and members indirectly bear the cost of the interest expense. In many bulk commodities such as grain, members can either sell at harvest or store the commodity for a fee until they decide to sell. Under that scenario, the members must also create their own marketing strategies.

Operating on a policy of COD is popular when producers have several marketing alternatives, especially when futures markets exist. This policy is more comparable with the operations of an IOF and therefore facilitates direct comparison between cooperatives and IOFs.

Delayed Payment

When a delayed payment policy is used, farmers deliver their commodities to the cooperative and may or may not retain title. Members may or may not receive an intermediate payment that represents a portion of the value of the commodity. The cooperative sells the commodities at the most advantageous prices. Expenses are then deducted, and the patron receives the residual amount. Delayed payment is desirable because the cooperative's exposure to risk and financial requirements is at a minimum, particularly if it does not take title and thus avoids the risks of ownership.

Moreover, the delayed payment structure is clearly in harmony with the business-at-cost principle because one year's business does not subsidize commodities marketed in a previous or subsequent year. It also substantially reduces working capital requirements of the cooperative because only partial payments are advanced to patrons before accounts have been closed. If members understand that they will indirectly be paying the interest expense for a COD system, they may support the delayed payment structure. Bankers and rating agencies also prefer delayed payments. It gives them added protection of default. Consequently, cooperatives using delayed payments can often receive borrowed funds at lower interest rates.

Many members, however, may not like the system because they do not know what they will receive until several months after delivery. Individuals, therefore, may be reluctant to join cooperatives that use delayed-payment strategies. The difficulties and costs of educating members about methods of delayed payment may also be a disadvantage. Delayed-payment practices include using individual accounts, auctions, and pools. In the following sections, we explain each of these practices.

Individual Account

If cooperatives establish individual accounts, deliveries by each patron are handled and traced separately, and products from other producers are not commingled. Cotton is marketed on an identity-preserved basis, with each producer receiving the price (minus expenses) associated with their specific bales. Marketing based on individual accounts or identity preservation can be desirable because producers receive what their individual products bring in the marketplace and because it comes nearest to complying with the business-at-cost principle. However, in many situations, this method is so unwieldy that it is not economically feasible. It is much more economical to handle and transport bulk commodities in large lots, which makes commingling necessary. Selling based on individual accounts also limits the market power since sales are fragmented.

Auctions

Auctions are another vehicle that farmers can use to sell their products through cooperatives and one that also retains individual accounts. Under the terms of auctions, payment is made shortly after the product is sold. Farmers may be attracted to auctions to take advantage of widely publicized competitive prices, low marketing costs, guaranteed payment, and reduced liability for marketing losses. There are some drawbacks to traditional auctions: Buyers must be physically present, and only a limited amount of the commodity can be sold per hour. For that reason, electronic auctions have increased in popularity but do not work for all commodities. Auctions are not well suited for large growers who find it advantageous to sell direct. Auctions are also less popular for storable or semidurable products such as grain because the commodity marketing structure with published bids and future market prices is more efficient.

Pooling

In a general sense, all cooperative activity pools risks, costs, and income. In the context of marketing cooperatives, pooling cooperatives commingle the products of their members and then sell the pool over the course of the marketing year. After all the commodities in the pool are marketed by the cooperative, expenses are deducted, and the average net price received is paid to producers. While an individual producer never receives the highest price that was possible over the year, it is also protected from selling at the market low. Key elements of a pool are the sharing of risks, expenses, revenues, and the payment of an average price, with possible adjustments for product quality and for time and location of delivery.

While each cooperative pool has its own operating procedures, most have the following characteristics. Farmers sign marketing contracts (see the following section) with the cooperative that guarantees delivery of all or part of their production to the pool. The contract transfers all authority over marketing decisions, including timing, pricing, and further processing, to the cooperative and its professional management. Members receive an initial advance payment upon delivery of the product. The advance is generally a percentage of an estimated market price. One or more progress payments may be made as the product is sold out of inventory.

When all or most of the product has been sold, generally 12 months or more after delivery, the pool is closed. A total value, which includes an estimated value of any remaining inventory, is determined for the pool. Operating and administrative expenses are allocated and subtracted. Any excess over previous payments is then distributed to patrons. Many producers consider the final payment to be patronage, although it is not possible to separate the commodity value from the cooperative's profits under the pool structure. A per-unit retain equity contribution is generally withheld from payments to producers. The per-unit retain is a category of revolving equity and ultimately redeemed into cash. The per-unit equity structure provides the equity capital necessary to finance the cooperative.

Milk, fruit, vegetable, and nut cooperatives generally use marketing pools. Sunkist Growers, Sun-Diamond Growers of California, Ocean Spray Cranberries, American Crystal Sugar Company, National Grape Cooperative Association, and California Almond Growers Exchange are cooperatives among the list of top 50 cooperatives that use marketing pools. Grain and oilseed marketing cooperatives have tried pools with mixed results. Pools are used in other countries such as Canada and Australia, to a greater extent than they are used in the United States. In recent years, larger cooperatives and marketing agencies in common (discussed in a subsequent section) have added marketing pools to the options offered to producers.

Marketing Contracts

Pooling cooperatives require marketing contracts. Contracts also are employed by other bargaining and marketing cooperatives. Marketing contracts (also called *member or marketing agreements*) are written agreements between the cooperative and members that state the rights and duties of both parties regarding how products will be marketed or purchased for a specified period of time. Contracts may in turn be tied to marketing pools, individual accounts, cash purchases, or bargaining cooperatives. The cooperative agrees to a certain payment method and other provisions necessary to market or acquire products to the best advantage of the members.

Agreement formalities between a cooperative and a patron range from nothing more than a purchase and a sale to an agreement binding the cooperative to provide a service and requiring the member to patronize the cooperative. Contractual arrangements require cooperatives to define carefully the relationship they desire with patrons and to spell out each party's rights and responsibilities under the agreement. Special consideration must be given to contracts because the rules governing cooperative-patron relationships are important and because contract-related provisions are contained in many cooperative statues.

Cooperatives may also develop contracts that involve a strict relationship between the member and the cooperative. These agreements are usually marketing contracts under which farmers agree to market all or a part of their specified commodities through the cooperative. Most cooperative statutes mention this type of contract, although special laws are not required for it. Most statutes simply state whether the cooperative may or may not take title to members' products and recognize the marketing and bargaining functions that cooperatives may perform. Many statutes also place a limit, usually 10 years, on the length of time for which a marketing contract can exist.

Typical Provisions

Cooperatives should incorporate clearly stated provisions into the contract or into bylaws. Typical provisions include the following:

1. The percentage or acreage of products the member must market through the cooperative.
2. An agreement that the cooperative will seek maximum net price for members.
3. The length of time the agreement is in force (e.g., one or three years). Conditions for renewal or termination often are included. The length of contract is largely determined by required investments and physical and marketing characteristics of the product. Commodities that are stored and marketed over a number of months or even years and that require substantial capital investments may require longer contracts.
4. The qualifications producers must meet for membership. First, they must qualify as farmers. In addition, they may have to qualify on the basis of their product, location, and ability to fulfill requirements of the marketing contract.
5. With increased sensitivity to food safety and environmental issues, many contracts spell out the cultural practices that farmers can and cannot use in growing their products. Authorization for per-unit capital retains to finance the cooperative. Per-unit capital retains are the prime source of equity capital. Many contracts include provisions for enforcement of the contract, penalties for noncompliance, and a description of damages to be paid to compensate for breach of the contract. Damages are an estimation, commonly called liquidated damages, of the loss suffered. Many statutes also make it illegal

for anyone to interfere with a cooperative's agreement with its patrons.
6. The method of payment for the product and control of product quality. Growing, harvesting, and delivery conditions may be specified to meet requirements of buyers and to minimize costs.
7. Control of quantity. This provision attempts to balance the quantity supplied with market requirements at prices that compensate growers for their inputs. The ability to specify these types of requirements is beneficial for both cooperatives and farmers. Increased use of specification buying has moved contracts to the forefront as a method of achieving coordination in meeting ultimate consumer needs.

Purpose of Contracts

Patrons receive many benefits when their cooperatives use contracts. Members like contracts because they ensure a market for their products. Contracts give producers an important advantage in securing credit, because loan officers recognize that a contract reduces risk and provides a market at reasonable prices. Further, enforced contracts help solve the free-rider problem.

Cooperatives can also improve their business by using contracts. They can reduce or eliminate costs associated with annual solicitation of patronage. More importantly, cooperatives can offer forward contracts with quantity, quality, and schedule guarantees and can acquire supplies and capital at more favorable rates and prices because of the image of reliability created by producer contracts.

Contracts enable cooperatives to plan for the best combination of resources to minimize costs. By using contracts, cooperatives can reduce procurement, assembly, and delivery costs because there are, for example, no solicitation costs. These savings may be offset by membership development, training, broken contracts, and legal fees. Contracts are of legal importance to both cooperatives and patrons because they restrain outside interference, guarantee that either side can collect for damages incurred by a breach of contract, and record the rights and duties of both cooperatives and their members.

Problems with Marketing Contracts

Contracts are not, however, appropriate for all situations, and even when they are appropriate, cooperatives need to make efforts to minimize their potential weaknesses. For example, some potential members may be hesitant to make the type of binding commitment that the contract requires. One of the greatest drawbacks for farmers who sign contracts, particularly for those producing bulk commodities, such as grain and oilseeds, is the loss of freedom to use other marketing alternatives. Contracts cannot be relied on simply because they can be enforced by the courts. Even IOFs are reluctant to sue farmers for breach of contract because of the invariable loss of goodwill that results from lawsuits. Many cooperatives are reluctant to take legal action against their member owners when contractual obligations are not met. However, failure to do so may result in costs to the overall membership.

Foreign Sourcing

There are often good business reasons for a cooperative to seek supplies from other countries to augment those provided by their members. Sourcing from another hemisphere may allow them to provide a year-round supply.

Operating on a global level can also increase their marketing power and give them greater ability to negotiate favorable prices. Cooperative members often have a knee-jerk reaction against foreign sourcing. That reaction can backfire when competing U.S. IOFs pursue the global opportunities.

The least controversial outsourcing situation is where a cooperative marketing perishable fruits and vegetables seeks off-season supplies so that they can offer their customers a constant, year-round flow of fresh products. This practice allows the cooperative to acquire year-round contracts. Two cooperatives in California have producer-members in foreign countries for this purpose. Foreign sourcing can create membership issues. If the foreign production is sourced as nonmember business, the cooperative stands the risk of exceeding 50% nonmember business and losing its cooperative status. However, if the cooperative has a one-member-one-vote system and the foreign producers are small-scale, there is a danger that adding the foreign producers as members will allow the foreign producers to control the cooperative.

Marketing Agencies in Common

Marketing agencies in common (MACs) refer to a cooperative formed to combine the marketing activities of multiple cooperatives. The MAC structure allows the individual cooperatives to achieve the economies of scale of a combined cooperative while maintaining the autonomy of the local cooperative structure. The activities and structures vary across MACs. CoMark Equity Alliance (CEA), a grain marketing alliance for Oklahoma and Kansas cooperatives, provides an example of the possible rationale. CEA markets grain for more than 140 cooperative locations. Prior to the formation of CEA (and its previous forms as two separate alliances), each cooperative had to employ a grain merchandiser. One professional grain merchandizer now performs those activities for the entire alliance. The alliance is also able to access additional market opportunities because of its size. A typical flour mill may require 1 million bushels of grain per week. A flour mill buyer is not interested in contacting individual cooperatives that may or may not have grain available to sell. Because of its size, the alliance always has grain available and can constantly provide bids to the flour mill. The grain alliance also operates under a single grain license. Because grain is owned by the farmer until it is sold, the grain handler must always maintain the grain inventory, matching the farmer-owned grain. Operating under one license, the alliance can hold that inventory anywhere in the system. This allows for more efficient use of storage space. If farmers anywhere in the system are holding grain going into harvest, that volume can be held in a location that does not generally fill up. Without the alliance, an individual cooperative would have to rent storage space at a regional elevator or risk not having sufficient space for harvest receipts. The single grain license also can be used to minimize transportation costs. When producers anywhere in the system sell grain and the best market is to the north, the alliance can move grain out of the northern elevator. Later, when other producers sell grain and the best market is in the South, grain can be moved out of southern elevators. Over the course of the marketing year, the alliance can market grain with lower transportation costs than those that the individual cooperatives would have faced.

As opposed to federated cooperatives, in which the local cooperatives have free choice as to whether to sell each lot of grain through a regional cooperative or a competitor, a MAC typically involves a contractual relationship mandating that all marketing flow to the alliance. This dedicated volume also contributes the alliance's ability to operate efficiently and create market power.

Several variations in the variety of marketing functions performed by MACs exist. Sun-Diamond, for example, whose five members market walnuts, raisins, prunes, figs, and hazelnuts, handles logistical, governmental relations, and international and nonmember business for its members. Its members retain other marketing functions of pricing and promotion and therefore bear all business risk. Their products retain independent identity in the market place. Some dairy MACs negotiate for price premiums over milk MO prices.

MACs are also used to spread the otherwise astronomical costs of market development to establish consumer-recognizable brand names. Economies of mass advertising are well established. Benefiting from the Norbest brand of turkeys, for example, are two independent turkey processing cooperatives. Member cooperatives sign a contract with Norbest, Inc., to use Norbest's marketing services and brand name. Contracts between a MAC and its members are much like those used between producers and cooperatives. Typically, members are required to sell all their products through the MAC. Per-unit capital retains or marketing fees are major sources of equity capital. Pooling and individual accounts may be used.

In spite of enormous potential payoff, the formation and continued operation of MACs face several challenges. Developing uniform policies on excess inventories, quality, packaging, and invoicing reaches back to some deeply imbedded practices by individual cooperative members. It isn't just the farmer-member who has to surrender prerogatives.

Value-Added Marketing

Any economic activity presumably adds value. *Value-added* is a continuum from near zero to highly processed and prepared food items. However, the term generally refers to processing a raw commodity into a more desirable intermediate product, such as Land O'Lakes butter, Ocean Spray juice, and Farmland processed meats. Farmers investing in new-generation cooperatives see themselves in the food manufacturing business rather than producing commodities more so than farmers who do not invest in these cooperatives. Participation in value-added cooperatives requires more up-front capital and corresponding risk. A wholly different level of management expertise and board ability are required. Many traditional boards are threatened by processing and consumer marketing because of their lack of exposure.

Consumer Product Marketing

An issue related to breadth of product line is forward integration, meaning the extent to which cooperatives should take products closer to the consumer. Several U.S. cooperatives are involved in marketing consumer or branded products. Some of these organizations started as bargaining cooperatives but over time evolved into vertically integrated organizations that sell branded products to food wholesalers, retailers, and institutions.

The general marketing strategies of these cooperatives are no different from those of other cooperatives or even IOFs. However, cooperatives that market consumer products face several additional challenges. A few of these challenges are discussed in the following sections. They include (1) developing an effective customer-oriented organization, (2) determining the appropriate role of new-product development in the product life cycle, and (3) establishing a strong and well-respected brand name.

A Customer Orientation

To be successful at marketing valued added products, a cooperative must have a consumer orientation. Doing it well requires experienced, creative, and customer-oriented marketing personnel who often command salaries higher than many cooperatives are accustomed to paying. Although the process sounds simple, it often is more difficult for cooperatives than for other types of organizations, because cooperatives place primary emphasis on satisfying their members as owners rather than customers. Consequently, there are likely to be conflicts that will arise in both supply and service cooperatives as well as marketing cooperatives. Cooperative members are often risk-takers but not risk-seekers.

For example, Ocean Spray discovered that consumers prefer a mix of cranberry juices and other juices. The development of a new juice line would have been simple in an IOF but was more controversial since the cooperative's members were cranberry growers and focused primarily on marketing cranberry juice. In other cases, consumer demand has dictated a change in the commodity supplied. For example, consumers may be concerned about genetically modified crops or pesticides. From a marketing perspective, it may be logical to change the supply chain to eliminate those issues. Some cooperative members may not understand why the cooperative is dictating how members produce their commodities. As a result, it is often difficult to please customers without losing crucial member support.

Another conflict sometimes arises when a cooperative limits the amount of a product it will take from its members. Most brand-name cooperative products have an extremely high-quality image, allowing the cooperative to charge premium prices. But the cooperative cannot charge premium prices if it floods the market with its products. This means that the cooperative may need to allocate and limit production to meet market demand. Value-added cooperatives have the unique challenge of attempting to always match demand with production.

New Products and Brand Development

Value-added marketing can require the development of new products and development and maintenance of brands. These activities are capital-intensive and require a long-term focus. Although no comparative studies are available, observation suggests that in general agricultural cooperatives are not overly aggressive in developing new products and markets, although there are some outstanding exceptions. Moreover, success is not guaranteed. Although many estimates exist, let us suppose that only three of 10 new products are likely to become profitable. If a cooperative is active in new product development, it will bear the costs of developing and testing all 10 new products, but only three will contribute to earnings. This means that there is significant risk involved, especially for agricultural markets that have relatively low profit margins.

Brand Development

To market consumer products successfully, cooperatives also need to develop and use a recognized and well-respected brand name. Several well-known cooperative brand names are Sunkist, Welch, Tree Top, Blue Diamond, Norbest, Land O'Lakes, and Ocean Spray. Some cooperatives capitalize on consumers' preference for products manufactured by farmer-owned businesses (see Figure 15.3) and incorporate the idea into their brand name. Gold Kist's adoption of the Farms brand and American Crystal Sugar Company's test marketing under the Farmer Direct label are examples. Despite these successes, agricultural cooperatives often do not have the capital or expertise to compete in the branded product marketplace.

Summary

Marketing cooperatives tend to be more specialized and focus on a single commodity. Marketing cooperatives use various structures involving different levels of commitment by their farmer members. Information-sharing cooperatives concentrate on monitoring and sharing market information that allows their producers to market more effectively. Bargaining cooperatives negotiate price and terms of trade but may not physically handle or take title to the commodity. Cooperatives that do purchase commodities from members and then resell the product can use various structures, including COD, auction, delayed payment, or marketing pool.

Federal MOs are not unique to cooperatives but are commonly associated with bargaining and marketing cooperatives. MOs can be designed to control quantity, quality, or marketing practices. If an MO is shown to be useful and economically justified, once approved by two-thirds of the producers in an area, it becomes legally binding on all producers. Many marketing cooperatives develop contracts with their members. Those contracts allow the cooperative to reduce risk, operate more efficiently, and increase market power. Although potentially beneficial to members, contracts can be problematic. Many cooperatives are reluctant to take legal actions with members to enforce marketing contracts.

Some marketing cooperatives source products from foreign sources. This can make good economic sense but be viewed unfavorably by members. It also raises the issue of nonmember business (if the foreign producers are not added as members) or loss of control of the cooperative (if the foreign producers are added as members). MACs involve centralizing the marketing activities of multiple cooperatives into a new cooperative. MACs can reduce costs through economies of scale and improve market access through volume and expertise. As opposed to the federated cooperative relationship, which is voluntarily conducted on a transaction-by-transaction basis, most MACs require member cooperatives to use the MACs exclusively for marketing.

Some marketing cooperatives pursue value-added and branded products. Those efforts are capital-intensive and involve additional risks as well as the possibility of higher returns. Value-added marketing often requires the cooperative to shift from a commodity focus to a consumer focus. This can lead to challenges when consumer preference have implication on the ingredient mix and/or commodity production process. There are well-known and successful cooperatives that market branded products. In general, cooperatives have had less presence in value-added and branded products then in bulk commodities due to the capital and expertise required, the risk, and the long time horizon of product development and payoff.

CHAPTER 18

Adjustments by Cooperatives

The previous chapters described the unique features of the cooperative business model and how that structure affects cooperative management, governance, strategy, and member benefit. While agricultural cooperatives are truly unique, one should not lose sight of the fact that they are progressive, profit-driven agribusiness firms operating in competitive environments. Agricultural cooperatives are constantly changing to meet the needs of their members and remain competitive in the marketplace. It is therefore difficult to make definitive statements about how agricultural cooperatives are adapting and evolving. It is clear that the agricultural cooperative industry is not static, and some major trends can be identified. We will begin by discussing adaptations of mature cooperatives.

Continued Search for Scale Economies

Agricultural cooperatives are in some dimensions an extension of their members' farm firms. Members of cooperatives are both users and owners. Members are affected by both prices on transactions and profits at the cooperative level. Because of that structure, agricultural cooperatives are particularly focused on efficiency. Increasing profits through efficiency is the win-win solution that allows the cooperative to prosper while not disadvantaging the member. That motivation explains one clear trend among agricultural cooperatives, which is a continued search for economies of scale.

Mergers and Unifications

In 2020, there were approximately 1,700 agricultural cooperatives in the United States, a 16% decline from the 2,310 cooperatives reported in 2010. During the same period, the net business volume of agricultural cooperatives increased from $146 billion to $200 billion. These statistics illustrate a historic and continuing trend of mergers and unifications among the cooperative industry. The market share of agricultural cooperatives has been stable or increased, but the number of cooperative firms has decreased.

Consolidation in the agricultural cooperative industry has made the terms *local cooperative* and *regional cooperative* less meaningful. Historically, a local cooperative was a single-location firm owned by the local producers. The term *regional cooperative* was applied to federated cooperatives owned by local cooperatives or by very large cooperatives operating over a multistate or national area. The federated regional structure created unique issues in regards to patronage, equity, governance, and accounting and was chiefly responsible for the discussions of regional cooperatives. Due to cooperative consolidations, many firms that would once have been described as local cooperatives now operate across entire states, and cross-state mergers are becoming increasingly common. The term *super local* is occasionally used to describe those firms, but because they are centralized (i.e., directly owned by producer-members), there has been little need to create a category for these multiple-location cooperatives.

The search for economies of scale and efficiency is clearly one of the driving forces behind the long-term trend of cooperative mergers. These unions can create numerous cost savings. Management and administrative functions are combined, which can bring substantial cost savings. Mergers also allow cooperatives to share equipment, combine functions, and operate closer to their peak capacity. Unification can also allow two cooperatives to capitalize on their comparative advantages. For example, a cooperative that excels in grain handling can combine with one that excels in agronomy and, in doing so, benefit both memberships.

Cooperative mergers are also often driven by management and human resource issues. A cooperative board that is facing the retirement of their long-time manager may explore merging with an adjacent cooperative with a strong, successful manager. Smaller cooperatives also have more difficulty attracting and retaining quality employees due to the lack of advancement opportunities within the organization. While existing employees are often skeptical of the benefits of a merger, over time, the larger, merged organization will likely have an easier time recruiting talent.

The changing structures of farming and the overall agribusiness industry have also contributed and will continue to contribute to mergers among cooperatives. Farm sizes continue to increase, and cooperative members now operate over a larger geographic area. This makes it more logical for adjacent cooperatives to combine operations and reduces members' knee-jerk reactions regarding the need to preserve local identity. Mergers between cooperatives in different geographic areas are becoming increasingly common. Combining nonadjacent cooperatives reduces exposure to weather risk.

The ongoing trend of consolidation among cooperatives has not occurred in isolation. All aspects of the agribusiness industry have become increasingly concentrated. Even after consolidation, agricultural cooperatives find themselves competing with much larger firms. Supply cooperatives find themselves contending with multinational corporations that sell directly to producers. Grain cooperatives compete with multinational firms that export to international markets. As they compete in a marketplace of giants, cooperatives face the need to increase their size and scope.

Joint Ventures and Alliances

Another strategy used by agricultural cooperatives to achieve scale and scope economies is the formation of joint ventures, alliances, and other multifirm structural relationships. In the early days of U.S. agricultural cooperatives, local cooperatives came together to form federated regional cooperatives as a vehicle for scale economies. Recent strategies involving joint ventures and alliances have the same goals but differ in several key aspects. These new entities are typically structured as limited liability companies (LLCs) rather than as new cooperatives. In some cases (e.g., grain marketing alliances), the ventures operate on a cooperative basis and are taxed as cooperatives. In other instances, the ventures are structured as investor-owned firms, and returns and voting are in proportion to ownership.

Unlike the federated regional structure, in which the local cooperative patronizes the regional on a voluntary basis, joint ventures and alliances usually involve strict contractual relationships. For example, in a grain marketing alliance, all the alliance members commit to marketing 100% of their grain through the alliance. In contrast with regional cooperatives that pursue their own strategies, most cooperative joint ventures and alliances are extremely focused. The ventures also often distribute all profits at the year's end and rely on capital calls to fund reinvestment or expansion.

Joint ventures, partnership, and alliances create new challenges for cooperative chief executive officers (CEOs) and boards of directors. A cooperative CEO may also be a board member of one or more ventures and thus have to take on oversight roles. The participants in the joint venture or alliance can also come to have different goals or a different assessment of performance. That can lead to disputes that must be navigated and, in some cases, exit strategies.

Innovations to Cooperative Finance

Base Capital

Some adjustments by agricultural cooperatives involve the cooperative financial model. As discussed in a previous chapter, closed-membership cooperatives, also termed *new-generation cooperatives*, couple equity with usage rights. This ensures that equity is kept in proportion to use and creates an incentive for equity investment. Hybrid member-investor cooperatives have a structure for accessing outside investment capital. Innovations involving equity among traditional open-membership cooperatives have been more modest but continue to occur.

Many regional and processing cooperatives have transitioned to a base capital system. In these cases, the logic of this type of system is easy to see. They are dealing with members, which, in some cases, are local cooperatives, that display constant or increasing business volume over time. This is in contrast with the lifetime business volume of farmer-members, who gradually increase patronage, decrease usage, and eventually exit the cooperative. The revolving-equity system, in which the cooperative is constantly creating and returning equity, is simply less logical for regional cooperatives.

The transition to base capital by regional cooperatives has in turn created new challenges for local cooperatives. Regional cooperatives distribute patronage in combination of cash and stock, and this patronage becomes part of the local cooperatives' total savings. Local cooperatives, in turn, distribute this total savings to their members in combinations of cash and stock patronage. This creates cash flow issues for the local cooperative if the cash/stock proportions vary between the local and regional and another set of cash flow issues if the equity redemption timing varies between the two levels. A local cooperative with a stable or growing business volume would never expect an equity redemption from a regional cooperative with a base capital system. This fact has caused some local cooperatives to explore new strategies for managing regional patronage. These include apportioning the regional equity to a new class of equity that would revolve to the member only when a redemption of regional equity occurred. Some local cooperatives retain regional equity patronage as unallocated retained earnings rather than reflecting it in their mix of cash and stock patronage.

Preferred Stock

Another financial innovation has been increasing use of preferred stock. Some regional cooperatives have distributed equity patronage in the form of interest-paying preferred stock. In some cases, the preferred shares are registered as securities and can be bought and sold by both members and outside investors. The preferred stock structure provides a benefit to holding the stock while at the same time creating semipermanent equity for the regional cooperative.

Some local cooperatives have innovated with preferred stock by converting revolving equity into dividend-bearing preferred stock after a given number of years. This decreases member dissatisfaction with the revolving period while allowing the cooperative to retain equity. The structure works best when the cooperative has nonmember profits. Dividends on capital stock are not deductible for most cooperatives, so paying dividends is not tax-efficient. Nonmember profits cannot be distributed as patronage and therefore create taxable income for the cooperative. Because the nonmember profits cannot be distributed in a tax-deductible form, there is no tax disadvantage to channeling them to stock dividends.

Unallocated Retained Earnings

Another trend in the financial structure of agricultural cooperatives (it is difficult to classify it as an innovation) has been increasing ratios of unallocated equity to allocated equity. In the early days of U.S. agricultural cooperatives, the vast majority of equity on the balance sheets was allocated equity (apportioned to specific members). Over time it became accepted that it was prudent for cooperatives to hold a small portion (5%–10%) of equity as unallocated retained earnings. Overtime, the ratio of unallocated equity to total equity increased for most cooperatives. The most dramatic change coincided with the implementation of the Section 199A tax deduction. While cooperatives can deduct profits distributed as allocated equity, retaining profits as unallocated retained earnings is not deductible and results in taxable income. The Section 199A deduction allowed cooperatives to offset that taxable income and therefore eliminated one deterrent to retaining member-based profits as unallocated equity. Many, if not most agricultural cooperatives have ratios of unallocated to total equity above 50%. A few local cooperatives have indicated strategies to transition to balance sheet structures where unallocated retained earnings accounts for the vast major of equity.

Nonqualified Stock

As discussed in previous chapters, cooperative can distribute equity patronage in the form of qualified stock which is immediately tax deductible to the cooperative and taxable to the member or nonqualified stock which is deductible and taxable in the future year when it is redeemed into cash. Local cooperatives have been slowly transitioning to nonqualified stock. The Tax Cuts and Jobs Act of 2017 reduced corporate tax rates and made nonqualified stock more logical. Regional cooperatives have also been transitioning to nonqualified stock patronage. That creates an issue for local cooperatives that calculate patronage on a tax basis. The regional nonqualified stock is not reflected as taxable income until the redemption year. That results in the pass-through patronage going to a different set of patrons from those who were using the cooperative when the regional patronage was earned. Local cooperatives distributing some or all of their stock patronage as nonqualified stock avoid that problem since they can issue nonqualified stock to their members to match the regional pass-through. This has been another factor in the slow transition toward nonqualified stock.

Strategic Innovations

The largest adjustment by agricultural cooperative is difficult to categorize. Simultaneous with efforts to gain economies of scale in traditional operations, most agricultural cooperatives have developed additional product lines, profit centers, and services. These adaptations are ongoing and diverse. The common denominator is that they allow the cooperative to increase profits and/or member benefits while taking advantage of marketplace opportunities. While it is impossible to list or even characterize the wide range of innovations, a few examples provide some insights.

Food processing cooperatives are constantly innovating to keep up with changing consumer demand. For example, Land O'Lakes recently introduced new queso products, butter balls, and blended cheese packages.

Innovations at Ocean Spray Cooperative include blended juices, Craisins (dried cranberries), snack mixes, energy drinks, and sparkling juices. MGB Marketing (a grower-owned blueberry cooperative) developed their Naturipe brand, which includes multiproduct snack products and yogurt parfaits. Numerous other examples exist, as product innovation is a constant process among processing cooperatives.

Farm supply cooperatives have diversified into seed treatment, soil sampling, precision agricultural services, mapping services, cover crop management, input financing, risk management, and numerous other products and services. Growmark Cooperative provides truck, rail, and barge logistic services and operates Growmark Energy, LLC, with 20 million gallons of petroleum storage, 150 bulk plant locations, and 300 tank wagons operating over a five-state area. The LLC also manufactures lubricants and diesel exhaust fluid. Tennessee Farmers Cooperative has expanded into farm and lifestyle products, outdoors and hunting equipment, turf and ornamental products, and show animal feed. The cooperative also operates a metal fabrication plant that produces livestock handling equipment. Southern States Cooperative offers heating, ventilation, and air conditioning consulting and service.

Grain marketing cooperatives have also actively diversified beyond their traditional commodity products. Central Valley Ag and Iowa Cooperative have numerous value-added grain programs, including white corn, non–genetically modified organism (GMO) yellow corn, non-GMO white corn, food-grade yellow corn, non-GMO waxy yellow corn, non-GMO high-amylose organic yellow corn, organic white corn, and organic soybeans. CHS, Inc., formed Ventura Foods, LLC, which is the nation's leading manufacturer and marketer of branded and custom-made shortenings, oils, dressing, sauces, mayonnaise, margarines culinary bases, and pan coatings. CHS also produces crude and refined canola oil and ethanol, processes packages, and delivers 10 varieties of dry edible beans. CHS is also involved with Arden Mills, a flour milling company and joint venture between CHS, Cargill, and Conagra Brands, Inc. Oklahoma grain cooperatives have explored niche opportunities for processing and marketing rye, capitalizing on the fact that Oklahoma is one of the few states in which rye is cultivated in a significant volume.

These examples, while admittedly random, give a glimpse at how agricultural cooperatives are evolving and adapting. Almost all agricultural cooperatives have established new activities during the past 20 years and will both add and eliminate activities over the next 20 years. Most agricultural cooperatives innovate or diversify into products and services with synergies with their existing operations and expertise. Innovations are also in response to marketplace opportunities, changing member needs, and the profile of members and potential members in their trade territories.

Challenges for Agricultural Cooperatives

Agricultural cooperatives face the same competitive challenges as investor-owned agribusinesses. However, the unique features of the cooperative business model create special challenges in adjusting to the changing business environment. Those challenges include capitalization, profit generation, communication and maintenance of democratic processes, and member loyalty and participation.

Capitalization Challenges

Unlike investor-owned firms, which can raise capital at any time, agricultural cooperatives create equity out of the profit stream. Their growth rate is limited by the level of their internally generated funds or, more precisely, by the portion of internally generated funds that they do not distribute to patrons. That creates an inherent challenge in funding the replacement and expansion of existing infrastructure as well as the capital required for new enterprises. The capital needs of the cooperative are always in competition with member expectations for cash patronage and retirement of previously issued equity. Financing the infrastructure to meet the needs of future generations of producers will continue to be a challenge for agricultural cooperatives.

Profit Generation

Although it seems too obvious to mention, an ongoing challenge for agricultural cooperatives is generating sufficient profits. The agribusiness marketplace is intensely competitive with large multinational firms that increasingly market directly to producers. In the period in which many of today's agricultural cooperatives were formed, producers clearly perceived the benefits of cooperatives in keeping markets competitive. Today's producers and cooperative members are much more transactional and expect the cooperative to have the most favorable price on every transaction. That creates an obvious challenge for the cooperative to maintain profitability. That profit is essential for the cooperative to create equity and fund infrastructure reinvestment. Identifying and maintaining profitable enterprises will be a continuing challenge for agricultural cooperatives.

Maintaining Democratic Processes

Cooperatives are member-owned, member-controlled organizations and can adapt and prosper in almost any business environment if the membership is unified behind their purpose. The membership of most agricultural cooperatives is now extremely heterogeneous, consisting of producers who have different needs and levels of commitment to the cooperatives. Cooperatives also face extremely competitive business environments that require them to anticipate and react to change. Agricultural cooperatives face ongoing challenges in making rapid decisions and executing their strategic plans while maintaining their democratic processes. Member communications and member involvement will remain critical challenges. Cooperative consolidation has increased those challenges. When a cooperative operates over a large geographical area, it is more difficult for its members to attend the annual meeting. Members are also less likely to be able to stop by the main office and visit with the CEO or a board member when the cooperative encompasses a large geographical region.

Cooperatives can and are using electronic communication and member engagement. They still face ongoing challenges in keeping members informed about the challenges and issues facing the cooperative and the rationale for the board's decisions. The challenge for mature cooperatives is that the majority of the members have never experienced a time when the cooperative was not active in the marketplace. Members evaluate the cooperative on transactions, and most factor in the possibility of patronage. Most are satisfied with their relationship with the cooperative but are not interested in deeper engagement. Other than electing board members, members in a cooperative actually vote on few issues. The major vehicle of member control has always been the members' ability to voice opinions to the board and CEO. Challenges emerge when the cooperative's rational business decisions adversely affect some segments of the membership. There is always the danger that instead of attempting to influence the cooperative's decisions, some members will simply exit it.

Adjustments by Newly Formed Cooperatives

The previous discussion has highlighted some of the adjustments and challenges of mature agricultural cooperatives. Another adjustment that has been and will continue to occur in the agricultural sector is the formation of new cooperatives. In general, these new cooperatives have different structures and operate in different business sectors.

Local Food Cooperatives

Most local food cooperatives require a relatively small initial equity investment in the form of a membership fee and require additional annual fees which are treated as income. In the early days of food cooperatives, members were encouraged to work in the store in return for a member discount. More recently, most food cooperatives have hired professional management and paid staff. Most local food cooperatives have a commitment to sourcing and marketing locally grown products. Local food cooperatives are often involved in efforts to support their local communities and may participate in local food networks that include farmers' markets and community-sponsored agricultural and other efforts. They may also work with local producers on farm production processes and production planning. Local food cooperatives tend to have frequent communication with both their consumer-members and their producer-suppliers. They also tend to have very committed and engaged membership, many of who have both financial and nonfinancial goals for the cooperative. Local food cooperatives are constantly innovating to maintain a competitive advantage in the procurement and promotion of locally grown foods.

Multistakeholder Cooperatives

Most cooperatives have membership that constitutes a single class of stakeholders. Producer cooperatives are owned by agricultural producers, consumer cooperatives are owned by consumers, and worker cooperatives are owned by workers. Multistakeholder cooperatives are owned and controlled by more than one type of membership class. Their membership may include consumers, workers, producers, volunteers, and community supporters. The stakeholders can be individuals, organizations such as nonprofits, businesses, government agencies, or even other cooperatives. Multistakeholder cooperatives tend to have broad missions, which reflects the interdependence of interest of multiple partners. They may be primarily focused on social objectives.

While multistakeholder cooperatives are a somewhat recent development, they are one of the fastest-growing segments of cooperatives. Multistakeholder cooperatives are challenged to provide tangible, meaningful benefits to members in the short term while trying to influence overall market structures in the long term. Another challenge is that with so many voices to be heard, the organization can become overburdened with governance tasks. On the other hand, when membership is broadened, larger issues such as unemployment, rural outmigration, environmental quality, member involvement, participation, and trust can actually increase.

Special-Purpose Cooperatives

New cooperatives are also formed for specific special needs or opportunities. Incubator cooperatives are focuses on helping small-scale beginning farmers develop viable farming operations. Machinery- and labor-sharing cooperatives help producers reduce costs and achieve economies of scale in machinery operation. There have also been examples of cooperative land ownership and community-owned farming operations.

Summary of New Cooperative Development and Adaptation

While representing a small portion of the agricultural cooperative industry, the greatest diversity in adaptation is occurring within newly established cooperatives. Those cooperatives provide ongoing examples of adjustment and innovation in both cooperative structure and strategy and purpose. Many of the groups forming these new organizations have complex objectives that involve financial, social, community, and marketplace goals. They have freely experimented with new structures to achieve these aims. While cooperatives, like other businesses, are difficult to launch, numerous studies have shown that once established, cooperatives last longer than other forms of businesses and are more resilient in times of crisis.

www.ingramcontent.com/pod-product-compliance
Ingram Content Group UK Ltd.
Pitfield, Milton Keynes, MK11 3LW, UK
UKHW061702190726
13853UKWH00008B/2357